C-2627 CAREER EXAMINATION SERIES

This is your PASSBOOK for...

Traffic Supervisor

Test Preparation Study Guide Questions & Answers

NLC®

NATIONAL LEARNING CORPORATION®

National Learning Corporation

212 Michael Drive, Syosset, NY 11791
(516) 921-8888 • www.passbooks.com
E-mail: info@passbooks.com

PUBLISHED IN THE UNITED STATES OF AMERICA

PASSBOOK® SERIES

THE *PASSBOOK® SERIES* has been created to prepare applicants and candidates for the ultimate academic battlefield – the examination room.

At some time in our lives, each and every one of us may be required to take an examination – for validation, matriculation, admission, qualification, registration, certification, or licensure.

Based on the assumption that every applicant or candidate has met the basic formal educational standards, has taken the required number of courses, and read the necessary texts, the *PASSBOOK® SERIES* furnishes the one special preparation which may assure passing with confidence, instead of failing with insecurity. Examination questions – together with answers – are furnished as the basic vehicle for study so that the mysteries of the examination and its compounding difficulties may be eliminated or diminished by a sure method.

This book is meant to help you pass your examination provided that you qualify and are serious in your objective.

The entire field is reviewed through the huge store of content information which is succinctly presented through a provocative and challenging approach – the question-and-answer method.

A climate of success is established by furnishing the correct answers at the end of each test.

You soon learn to recognize types of questions, forms of questions, and patterns of questioning. You may even begin to anticipate expected outcomes.

You perceive that many questions are repeated or adapted so that you can gain acute insights, which may enable you to score many sure points.

You learn how to confront new questions, or types of questions, and to attack them confidently and work out the correct answers.

You note objectives and emphases, and recognize pitfalls and dangers, so that you may make positive educational adjustments.

Moreover, you are kept fully informed in relation to new concepts, methods, practices, and directions in the field.

You discover that you are actually taking the examination all the time: you are preparing for the examination by "taking" an examination, not by reading extraneous and/or supererogatory textbooks.

In short, this PASSBOOK®, used directedly, should be an important factor in helping you to pass your test.

TRAFFIC SUPERVISOR

DUTIES:

As a Traffic Supervisor, you would plan, direct, review, and evaluate all activities relating to the administration of a Traffic Management Program within a Thruway division. In addition, you would supervise Division Traffic Personnel, prepare the division program budget and, if necessary, perform the duties of the Assistant Traffic Supervisor:

SUBJECT OF EXAMINATION:

The written test is designed to test for knowledge, skills, and/or abilities in such areas as:

1. **Principles and practices of traffic and transportation technology** - These questions test for knowledge of traffic and transportation technology concepts and their practical applications, including materials, terminology and procedures used, computations related to the construction and operation of roadways and related facilities, traffic and transportation studies and data interpretation, and safety issues.
2. **Traffic control devices and regulations and collection, analysis and presentation of data** - These questions test for knowledge of the uses, placement, laws, rules and regulations related to various types of traffic safety and traffic control devices such as pavement markings, work zone markings, signs and traffic signals, and the proper procedures and terminology used to gather, evaluate, organize and utilize various types of technical data related to transportation and traffic studies.
3. **Plans and specifications** - These questions test for the ability to understand, analyze, and perform computations based on various roadway-related technical drawings and written technical material. All the information required to answer the questions will be presented in the written material and/or drawings.
4. **Safety practices** - These questions test for knowledge of, and the ability to apply, safety principles related to highway work zones, including traffic control, the safe use of equipment, and the overall safety of workers, the traveling public and the work environment.
5. **Traffic accident investigative techniques including court testimony** - These questions test for knowledge of fundamental concepts relating to, and the procedures used in, the investigation of traffic accidents. The questions will deal with such areas as obtaining the cooperation of individuals involved in an investigation, preparing for the investigation and/or interview, investigative and interviewing principles and practices, gathering and evaluating data and evidence, presenting the results of an investigation, preparing a case for a court hearing, testifying in the courtroom or in hearings, and familiarity with basic legal terminology. Many of the questions are situational in nature and attempt to measure the candidates' ability to apply basic investigative techniques.
6. **Preparing written material** - These questions test for the ability to present information clearly and accurately, and to organize paragraphs logically and comprehensibly. For some questions, you will be given information in two or three sentences followed by four restatements of the information. You must then choose the best version. For other questions, you will be given paragraphs with their sentences out of order. You must then choose, from four suggestions, the best order for the sentences.
7. **Supervision** - These questions test for knowledge of the principles and practices employed in planning, organizing, and controlling the activities of a work unit toward predetermined objectives. The concepts covered, usually in a situational question format, include such topics as assigning and reviewing work; evaluating performance; maintaining work standards; motivating and developing subordinates; implementing procedural change; increasing efficiency; and dealing with problems of absenteeism, morale, and discipline.

HOW TO TAKE A TEST

I. YOU MUST PASS AN EXAMINATION

A. WHAT EVERY CANDIDATE SHOULD KNOW

Examination applicants often ask us for help in preparing for the written test. What can I study in advance? What kinds of questions will be asked? How will the test be given? How will the papers be graded?

As an applicant for a civil service examination, you may be wondering about some of these things. Our purpose here is to suggest effective methods of advance study and to describe civil service examinations.

Your chances for success on this examination can be increased if you know how to prepare. Those "pre-examination jitters" can be reduced if you know what to expect. You can even experience an adventure in good citizenship if you know why civil service exams are given.

B. WHY ARE CIVIL SERVICE EXAMINATIONS GIVEN?

Civil service examinations are important to you in two ways. As a citizen, you want public jobs filled by employees who know how to do their work. As a job seeker, you want a fair chance to compete for that job on an equal footing with other candidates. The best-known means of accomplishing this two-fold goal is the competitive examination.

Exams are widely publicized throughout the nation. They may be administered for jobs in federal, state, city, municipal, town or village governments or agencies.

Any citizen may apply, with some limitations, such as the age or residence of applicants. Your experience and education may be reviewed to see whether you meet the requirements for the particular examination. When these requirements exist, they are reasonable and applied consistently to all applicants. Thus, a competitive examination may cause you some uneasiness now, but it is your privilege and safeguard.

C. HOW ARE CIVIL SERVICE EXAMS DEVELOPED?

Examinations are carefully written by trained technicians who are specialists in the field known as "psychological measurement," in consultation with recognized authorities in the field of work that the test will cover. These experts recommend the subject matter areas or skills to be tested; only those knowledges or skills important to your success on the job are included. The most reliable books and source materials available are used as references. Together, the experts and technicians judge the difficulty level of the questions.

Test technicians know how to phrase questions so that the problem is clearly stated. Their ethics do not permit "trick" or "catch" questions. Questions may have been tried out on sample groups, or subjected to statistical analysis, to determine their usefulness.

Written tests are often used in combination with performance tests, ratings of training and experience, and oral interviews. All of these measures combine to form the best-known means of finding the right person for the right job.

II. HOW TO PASS THE WRITTEN TEST

A. NATURE OF THE EXAMINATION

To prepare intelligently for civil service examinations, you should know how they differ from school examinations you have taken. In school you were assigned certain definite pages to read or subjects to cover. The examination questions were quite detailed and usually emphasized memory. Civil service exams, on the other hand, try to discover your present ability to perform the duties of a position, plus your potentiality to learn these duties. In other words, a civil service exam attempts to predict how successful you will be. Questions cover such a broad area that they cannot be as minute and detailed as school exam questions.

In the public service similar kinds of work, or positions, are grouped together in one "class." This process is known as *position-classification.* All the positions in a class are paid according to the salary range for that class. One class title covers all of these positions, and they are all tested by the same examination.

B. FOUR BASIC STEPS

1) Study the announcement

How, then, can you know what subjects to study? Our best answer is: "Learn as much as possible about the class of positions for which you've applied." The exam will test the knowledge, skills and abilities needed to do the work.

Your most valuable source of information about the position you want is the official exam announcement. This announcement lists the training and experience qualifications. Check these standards and apply only if you come reasonably close to meeting them.

The brief description of the position in the examination announcement offers some clues to the subjects which will be tested. Think about the job itself. Review the duties in your mind. Can you perform them, or are there some in which you are rusty? Fill in the blank spots in your preparation.

Many jurisdictions preview the written test in the exam announcement by including a section called "Knowledge and Abilities Required," "Scope of the Examination," or some similar heading. Here you will find out specifically what fields will be tested.

2) Review your own background

Once you learn in general what the position is all about, and what you need to know to do the work, ask yourself which subjects you already know fairly well and which need improvement. You may wonder whether to concentrate on improving your strong areas or on building some background in your fields of weakness. When the announcement has specified "some knowledge" or "considerable knowledge," or has used adjectives like "beginning principles of..." or "advanced ... methods," you can get a clue as to the number and difficulty of questions to be asked in any given field. More questions, and hence broader coverage, would be included for those subjects which are more important in the work. Now weigh your strengths and weaknesses against the job requirements and prepare accordingly.

3) Determine the level of the position

Another way to tell how intensively you should prepare is to understand the level of the job for which you are applying. Is it the entering level? In other words, is this the position in which beginners in a field of work are hired? Or is it an intermediate or advanced level? Sometimes this is indicated by such words as "Junior" or "Senior" in the class title. Other jurisdictions use Roman numerals to designate the level – Clerk I, Clerk II, for example. The word "Supervisor" sometimes appears in the title. If the level is not indicated by the title, check the description of duties. Will you be working under very close supervision, or will you have responsibility for independent decisions in this work?

4) Choose appropriate study materials

Now that you know the subjects to be examined and the relative amount of each subject to be covered, you can choose suitable study materials. For beginning level jobs, or even advanced ones, if you have a pronounced weakness in some aspect of your training, read a modern, standard textbook in that field. Be sure it is up to date and has general coverage. Such books are normally available at your library, and the librarian will be glad to help you locate one. For entry-level positions, questions of appropriate difficulty are chosen – neither highly advanced questions, nor those too simple. Such questions require careful thought but not advanced training.

If the position for which you are applying is technical or advanced, you will read more advanced, specialized material. If you are already familiar with the basic principles of your field, elementary textbooks would waste your time. Concentrate on advanced textbooks and technical periodicals. Think through the concepts and review difficult problems in your field.

These are all general sources. You can get more ideas on your own initiative, following these leads. For example, training manuals and publications of the government agency which employs workers in your field can be useful, particularly for technical and professional positions. A letter or visit to the government department involved may result in more specific study suggestions, and certainly will provide you with a more definite idea of the exact nature of the position you are seeking.

III. KINDS OF TESTS

Tests are used for purposes other than measuring knowledge and ability to perform specified duties. For some positions, it is equally important to test ability to make adjustments to new situations or to profit from training. In others, basic mental abilities not dependent on information are essential. Questions which test these things may not appear as pertinent to the duties of the position as those which test for knowledge and information. Yet they are often highly important parts of a fair examination. For very general questions, it is almost impossible to help you direct your study efforts. What we can do is to point out some of the more common of these general abilities needed in public service positions and describe some typical questions.

1) General information

Broad, general information has been found useful for predicting job success in some kinds of work. This is tested in a variety of ways, from vocabulary lists to questions about current events. Basic background in some field of work, such as sociology or economics, may be sampled in a group of questions. Often these are principles which have become familiar to most persons through exposure rather than through formal training. It is difficult to advise you how to study for these questions; being alert to the world around you is our best suggestion.

2) Verbal ability

An example of an ability needed in many positions is verbal or language ability. Verbal ability is, in brief, the ability to use and understand words. Vocabulary and grammar tests are typical measures of this ability. Reading comprehension or paragraph interpretation questions are common in many kinds of civil service tests. You are given a paragraph of written material and asked to find its central meaning.

3) Numerical ability

Number skills can be tested by the familiar arithmetic problem, by checking paired lists of numbers to see which are alike and which are different, or by interpreting charts and graphs. In the latter test, a graph may be printed in the test booklet which you are asked to use as the basis for answering questions.

4) Observation

A popular test for law-enforcement positions is the observation test. A picture is shown to you for several minutes, then taken away. Questions about the picture test your ability to observe both details and larger elements.

5) Following directions

In many positions in the public service, the employee must be able to carry out written instructions dependably and accurately. You may be given a chart with several columns, each column listing a variety of information. The questions require you to carry out directions involving the information given in the chart.

6) Skills and aptitudes

Performance tests effectively measure some manual skills and aptitudes. When the skill is one in which you are trained, such as typing or shorthand, you can practice. These tests are often very much like those given in business school or high school courses. For many of the other skills and aptitudes, however, no short-time preparation can be made. Skills and abilities natural to you or that you have developed throughout your lifetime are being tested.

Many of the general questions just described provide all the data needed to answer the questions and ask you to use your reasoning ability to find the answers. Your best preparation for these tests, as well as for tests of facts and ideas, is to be at your physical and mental best. You, no doubt, have your own methods of getting into an exam-taking mood and keeping "in shape." The next section lists some ideas on this subject.

IV. KINDS OF QUESTIONS

Only rarely is the "essay" question, which you answer in narrative form, used in civil service tests. Civil service tests are usually of the short-answer type. Full instructions for answering these questions will be given to you at the examination. But in case this is your first experience with short-answer questions and separate answer sheets, here is what you need to know:

1) Multiple-choice Questions

Most popular of the short-answer questions is the "multiple choice" or "best answer" question. It can be used, for example, to test for factual knowledge, ability to solve problems or judgment in meeting situations found at work.

A multiple-choice question is normally one of three types—

- It can begin with an incomplete statement followed by several possible endings. You are to find the one ending which *best* completes the statement, although some of the others may not be entirely wrong.
- It can also be a complete statement in the form of a question which is answered by choosing one of the statements listed.

- It can be in the form of a problem – again you select the best answer.

Here is an example of a multiple-choice question with a discussion which should give you some clues as to the method for choosing the right answer:

When an employee has a complaint about his assignment, the action which will *best* help him overcome his difficulty is to
A. discuss his difficulty with his coworkers
B. take the problem to the head of the organization
C. take the problem to the person who gave him the assignment
D. say nothing to anyone about his complaint

In answering this question, you should study each of the choices to find which is best. Consider choice "A" – Certainly an employee may discuss his complaint with fellow employees, but no change or improvement can result, and the complaint remains unresolved. Choice "B" is a poor choice since the head of the organization probably does not know what assignment you have been given, and taking your problem to him is known as "going over the head" of the supervisor. The supervisor, or person who made the assignment, is the person who can clarify it or correct any injustice. Choice "C" is, therefore, correct. To say nothing, as in choice "D," is unwise. Supervisors have and interest in knowing the problems employees are facing, and the employee is seeking a solution to his problem.

2) True/False Questions

The "true/false" or "right/wrong" form of question is sometimes used. Here a complete statement is given. Your job is to decide whether the statement is right or wrong.

SAMPLE: A roaming cell-phone call to a nearby city costs less than a non-roaming call to a distant city.

This statement is wrong, or false, since roaming calls are more expensive.

This is not a complete list of all possible question forms, although most of the others are variations of these common types. You will always get complete directions for answering questions. Be sure you understand *how* to mark your answers – ask questions until you do.

V. RECORDING YOUR ANSWERS

Computer terminals are used more and more today for many different kinds of exams.

For an examination with very few applicants, you may be told to record your answers in the test booklet itself. Separate answer sheets are much more common. If this separate answer sheet is to be scored by machine – and this is often the case – it is highly important that you mark your answers correctly in order to get credit.

An electronic scoring machine is often used in civil service offices because of the speed with which papers can be scored. Machine-scored answer sheets must be marked with a pencil, which will be given to you. This pencil has a high graphite content which responds to the electronic scoring machine. As a matter of fact, stray dots may register as answers, so do not let your pencil rest on the answer sheet while you are pondering the correct answer. Also, if your pencil lead breaks or is otherwise defective, ask for another.

Since the answer sheet will be dropped in a slot in the scoring machine, be careful not to bend the corners or get the paper crumpled.

The answer sheet normally has five vertical columns of numbers, with 30 numbers to a column. These numbers correspond to the question numbers in your test booklet. After each number, going across the page are four or five pairs of dotted lines. These short dotted lines have small letters or numbers above them. The first two pairs may also have a "T" or "F" above the letters. This indicates that the first two pairs only are to be used if the questions are of the true-false type. If the questions are multiple choice, disregard the "T" and "F" and pay attention only to the small letters or numbers.

Answer your questions in the manner of the sample that follows:

32. The largest city in the United States is
    - A. Washington, D.C.
    - B. New York City
    - C. Chicago
    - D. Detroit
    - E. San Francisco

1) Choose the answer you think is best. (New York City is the largest, so "B" is correct.)
2) Find the row of dotted lines numbered the same as the question you are answering. (Find row number 32)
3) Find the pair of dotted lines corresponding to the answer. (Find the pair of lines under the mark "B.")
4) Make a solid black mark between the dotted lines.

VI. BEFORE THE TEST

Common sense will help you find procedures to follow to get ready for an examination. Too many of us, however, overlook these sensible measures. Indeed, nervousness and fatigue have been found to be the most serious reasons why applicants fail to do their best on civil service tests. Here is a list of reminders:

- Begin your preparation early – Don't wait until the last minute to go scurrying around for books and materials or to find out what the position is all about.
- Prepare continuously – An hour a night for a week is better than an all-night cram session. This has been definitely established. What is more, a night a week for a month will return better dividends than crowding your study into a shorter period of time.
- Locate the place of the exam – You have been sent a notice telling you when and where to report for the examination. If the location is in a different town or otherwise unfamiliar to you, it would be well to inquire the best route and learn something about the building.
- Relax the night before the test – Allow your mind to rest. Do not study at all that night. Plan some mild recreation or diversion; then go to bed early and get a good night's sleep.
- Get up early enough to make a leisurely trip to the place for the test – This way unforeseen events, traffic snarls, unfamiliar buildings, etc. will not upset you.
- Dress comfortably – A written test is not a fashion show. You will be known by number and not by name, so wear something comfortable.

- Leave excess paraphernalia at home – Shopping bags and odd bundles will get in your way. You need bring only the items mentioned in the official notice you received; usually everything you need is provided. Do not bring reference books to the exam. They will only confuse those last minutes and be taken away from you when in the test room.
- Arrive somewhat ahead of time – If because of transportation schedules you must get there very early, bring a newspaper or magazine to take your mind off yourself while waiting.
- Locate the examination room – When you have found the proper room, you will be directed to the seat or part of the room where you will sit. Sometimes you are given a sheet of instructions to read while you are waiting. Do not fill out any forms until you are told to do so; just read them and be prepared.
- Relax and prepare to listen to the instructions
- If you have any physical problem that may keep you from doing your best, be sure to tell the test administrator. If you are sick or in poor health, you really cannot do your best on the exam. You can come back and take the test some other time.

VII. AT THE TEST

The day of the test is here and you have the test booklet in your hand. The temptation to get going is very strong. Caution! There is more to success than knowing the right answers. You must know how to identify your papers and understand variations in the type of short-answer question used in this particular examination. Follow these suggestions for maximum results from your efforts:

1) Cooperate with the monitor

The test administrator has a duty to create a situation in which you can be as much at ease as possible. He will give instructions, tell you when to begin, check to see that you are marking your answer sheet correctly, and so on. He is not there to guard you, although he will see that your competitors do not take unfair advantage. He wants to help you do your best.

2) Listen to all instructions

Don't jump the gun! Wait until you understand all directions. In most civil service tests you get more time than you need to answer the questions. So don't be in a hurry. Read each word of instructions until you clearly understand the meaning. Study the examples, listen to all announcements and follow directions. Ask questions if you do not understand what to do.

3) Identify your papers

Civil service exams are usually identified by number only. You will be assigned a number; you must not put your name on your test papers. Be sure to copy your number correctly. Since more than one exam may be given, copy your exact examination title.

4) Plan your time

Unless you are told that a test is a "speed" or "rate of work" test, speed itself is usually not important. Time enough to answer all the questions will be provided, but this does not mean that you have all day. An overall time limit has been set. Divide the total time (in minutes) by the number of questions to determine the approximate time you have for each question.

5) Do not linger over difficult questions

If you come across a difficult question, mark it with a paper clip (useful to have along) and come back to it when you have been through the booklet. One caution if you do this – be sure to skip a number on your answer sheet as well. Check often to be sure that you have not lost your place and that you are marking in the row numbered the same as the question you are answering.

6) Read the questions

Be sure you know what the question asks! Many capable people are unsuccessful because they failed to *read* the questions correctly.

7) Answer all questions

Unless you have been instructed that a penalty will be deducted for incorrect answers, it is better to guess than to omit a question.

8) Speed tests

It is often better NOT to guess on speed tests. It has been found that on timed tests people are tempted to spend the last few seconds before time is called in marking answers at random – without even reading them – in the hope of picking up a few extra points. To discourage this practice, the instructions may warn you that your score will be "corrected" for guessing. That is, a penalty will be applied. The incorrect answers will be deducted from the correct ones, or some other penalty formula will be used.

9) Review your answers

If you finish before time is called, go back to the questions you guessed or omitted to give them further thought. Review other answers if you have time.

10) Return your test materials

If you are ready to leave before others have finished or time is called, take ALL your materials to the monitor and leave quietly. Never take any test material with you. The monitor can discover whose papers are not complete, and taking a test booklet may be grounds for disqualification.

VIII. EXAMINATION TECHNIQUES

1) Read the general instructions carefully. These are usually printed on the first page of the exam booklet. As a rule, these instructions refer to the timing of the examination; the fact that you should not start work until the signal and must stop work at a signal, etc. If there are any *special* instructions, such as a choice of questions to be answered, make sure that you note this instruction carefully.

2) When you are ready to start work on the examination, that is as soon as the signal has been given, read the instructions to each question booklet, underline any key words or phrases, such as *least, best, outline, describe* and the like. In this way you will tend to answer as requested rather than discover on reviewing your paper that you *listed without describing*, that you selected the *worst* choice rather than the *best* choice, etc.

3) If the examination is of the objective or multiple-choice type – that is, each question will also give a series of possible answers: A, B, C or D, and you are called upon to select the best answer and write the letter next to that answer on your answer paper – it is advisable to start answering each question in turn. There may be anywhere from 50 to 100 such questions in the three or four hours allotted and you can see how much time would be taken if you read through all the questions before beginning to answer any. Furthermore, if you come across a question or group of questions which you know would be difficult to answer, it would undoubtedly affect your handling of all the other questions.

4) If the examination is of the essay type and contains but a few questions, it is a moot point as to whether you should read all the questions before starting to answer any one. Of course, if you are given a choice – say five out of seven and the like – then it is essential to read all the questions so you can eliminate the two that are most difficult. If, however, you are asked to answer all the questions, there may be danger in trying to answer the easiest one first because you may find that you will spend too much time on it. The best technique is to answer the first question, then proceed to the second, etc.

5) Time your answers. Before the exam begins, write down the time it started, then add the time allowed for the examination and write down the time it must be completed, then divide the time available somewhat as follows:
   - If 3-1/2 hours are allowed, that would be 210 minutes. If you have 80 objective-type questions, that would be an average of 2-1/2 minutes per question. Allow yourself no more than 2 minutes per question, or a total of 160 minutes, which will permit about 50 minutes to review.
   - If for the time allotment of 210 minutes there are 7 essay questions to answer, that would average about 30 minutes a question. Give yourself only 25 minutes per question so that you have about 35 minutes to review.

6) The most important instruction is to *read each question* and make sure you know what is wanted. The second most important instruction is to *time yourself properly* so that you answer every question. The third most important instruction is to *answer every question.* Guess if you have to but include something for each question. Remember that you will receive no credit for a blank and will probably receive some credit if you write something in answer to an essay question. If you guess a letter – say "B" for a multiple-choice question – you may have guessed right. If you leave a blank as an answer to a multiple-choice question, the examiners may respect your feelings but it will not add a point to your score. Some exams may penalize you for wrong answers, so in such cases *only*, you may not want to guess unless you have some basis for your answer.

7) Suggestions
   a. Objective-type questions
      1. Examine the question booklet for proper sequence of pages and questions
      2. Read all instructions carefully
      3. Skip any question which seems too difficult; return to it after all other questions have been answered
      4. Apportion your time properly; do not spend too much time on any single question or group of questions

5. Note and underline key words – *all, most, fewest, least, best, worst, same, opposite,* etc.
6. Pay particular attention to negatives
7. Note unusual option, e.g., unduly long, short, complex, different or similar in content to the body of the question
8. Observe the use of "hedging" words – *probably, may, most likely,* etc.
9. Make sure that your answer is put next to the same number as the question
10. Do not second-guess unless you have good reason to believe the second answer is definitely more correct
11. Cross out original answer if you decide another answer is more accurate; do not erase until you are ready to hand your paper in
12. Answer all questions; guess unless instructed otherwise
13. Leave time for review

b. Essay questions

1. Read each question carefully
2. Determine exactly what is wanted. Underline key words or phrases.
3. Decide on outline or paragraph answer
4. Include many different points and elements unless asked to develop any one or two points or elements
5. Show impartiality by giving pros and cons unless directed to select one side only
6. Make and write down any assumptions you find necessary to answer the questions
7. Watch your English, grammar, punctuation and choice of words
8. Time your answers; don't crowd material

8) Answering the essay question

Most essay questions can be answered by framing the specific response around several key words or ideas. Here are a few such key words or ideas:

M's: manpower, materials, methods, money, management

P's: purpose, program, policy, plan, procedure, practice, problems, pitfalls, personnel, public relations

a. Six basic steps in handling problems:

1. Preliminary plan and background development
2. Collect information, data and facts
3. Analyze and interpret information, data and facts
4. Analyze and develop solutions as well as make recommendations
5. Prepare report and sell recommendations
6. Install recommendations and follow up effectiveness

b. Pitfalls to avoid

1. *Taking things for granted* – A statement of the situation does not necessarily imply that each of the elements is necessarily true; for example, a complaint may be invalid and biased so that all that can be taken for granted is that a complaint has been registered

2. *Considering only one side of a situation* – Wherever possible, indicate several alternatives and then point out the reasons you selected the best one
3. *Failing to indicate follow up* – Whenever your answer indicates action on your part, make certain that you will take proper follow-up action to see how successful your recommendations, procedures or actions turn out to be
4. *Taking too long in answering any single question* – Remember to time your answers properly

IX. AFTER THE TEST

Scoring procedures differ in detail among civil service jurisdictions although the general principles are the same. Whether the papers are hand-scored or graded by machine we have described, they are nearly always graded by number. That is, the person who marks the paper knows only the number – never the name – of the applicant. Not until all the papers have been graded will they be matched with names. If other tests, such as training and experience or oral interview ratings have been given, scores will be combined. Different parts of the examination usually have different weights. For example, the written test might count 60 percent of the final grade, and a rating of training and experience 40 percent. In many jurisdictions, veterans will have a certain number of points added to their grades.

After the final grade has been determined, the names are placed in grade order and an eligible list is established. There are various methods for resolving ties between those who get the same final grade – probably the most common is to place first the name of the person whose application was received first. Job offers are made from the eligible list in the order the names appear on it. You will be notified of your grade and your rank as soon as all these computations have been made. This will be done as rapidly as possible.

People who are found to meet the requirements in the announcement are called "eligibles." Their names are put on a list of eligible candidates. An eligible's chances of getting a job depend on how high he stands on this list and how fast agencies are filling jobs from the list.

When a job is to be filled from a list of eligibles, the agency asks for the names of people on the list of eligibles for that job. When the civil service commission receives this request, it sends to the agency the names of the three people highest on this list. Or, if the job to be filled has specialized requirements, the office sends the agency the names of the top three persons who meet these requirements from the general list.

The appointing officer makes a choice from among the three people whose names were sent to him. If the selected person accepts the appointment, the names of the others are put back on the list to be considered for future openings.

That is the rule in hiring from all kinds of eligible lists, whether they are for typist, carpenter, chemist, or something else. For every vacancy, the appointing officer has his choice of any one of the top three eligibles on the list. This explains why the person whose name is on top of the list sometimes does not get an appointment when some of the persons lower on the list do. If the appointing officer chooses the second or third eligible, the No. 1 eligible does not get a job at once, but stays on the list until he is appointed or the list is terminated.

X. HOW TO PASS THE INTERVIEW TEST

The examination for which you applied requires an oral interview test. You have already taken the written test and you are now being called for the interview test – the final part of the formal examination.

You may think that it is not possible to prepare for an interview test and that there are no procedures to follow during an interview. Our purpose is to point out some things you can do in advance that will help you and some good rules to follow and pitfalls to avoid while you are being interviewed.

What is an interview supposed to test?

The written examination is designed to test the technical knowledge and competence of the candidate; the oral is designed to evaluate intangible qualities, not readily measured otherwise, and to establish a list showing the relative fitness of each candidate – as measured against his competitors – for the position sought. Scoring is not on the basis of "right" and "wrong," but on a sliding scale of values ranging from "not passable" to "outstanding." As a matter of fact, it is possible to achieve a relatively low score without a single "incorrect" answer because of evident weakness in the qualities being measured.

Occasionally, an examination may consist entirely of an oral test – either an individual or a group oral. In such cases, information is sought concerning the technical knowledges and abilities of the candidate, since there has been no written examination for this purpose. More commonly, however, an oral test is used to supplement a written examination.

Who conducts interviews?

The composition of oral boards varies among different jurisdictions. In nearly all, a representative of the personnel department serves as chairman. One of the members of the board may be a representative of the department in which the candidate would work. In some cases, "outside experts" are used, and, frequently, a businessman or some other representative of the general public is asked to serve. Labor and management or other special groups may be represented. The aim is to secure the services of experts in the appropriate field.

However the board is composed, it is a good idea (and not at all improper or unethical) to ascertain in advance of the interview who the members are and what groups they represent. When you are introduced to them, you will have some idea of their backgrounds and interests, and at least you will not stutter and stammer over their names.

What should be done before the interview?

While knowledge about the board members is useful and takes some of the surprise element out of the interview, there is other preparation which is more substantive. It *is* possible to prepare for an oral interview – in several ways:

1) Keep a copy of your application and review it carefully before the interview

This may be the only document before the oral board, and the starting point of the interview. Know what education and experience you have listed there, and the sequence and dates of all of it. Sometimes the board will ask you to review the highlights of your experience for them; you should not have to hem and haw doing it.

2) Study the class specification and the examination announcement

Usually, the oral board has one or both of these to guide them. The qualities, characteristics or knowledges required by the position sought are stated in these documents. They offer valuable clues as to the nature of the oral interview. For example, if the job involves supervisory responsibilities, the announcement will usually indicate that knowledge of modern supervisory methods and the qualifications of the candidate as a supervisor will be tested. If so, you can expect such questions, frequently in the form of a hypothetical situation which you are expected to solve. NEVER go into an oral without knowledge of the duties and responsibilities of the job you seek.

3) Think through each qualification required

Try to visualize the kind of questions you would ask if you were a board member. How well could you answer them? Try especially to appraise your own knowledge and background in each area, *measured against the job sought*, and identify any areas in which you are weak. Be critical and realistic – do not flatter yourself.

4) Do some general reading in areas in which you feel you may be weak

For example, if the job involves supervision and your past experience has NOT, some general reading in supervisory methods and practices, particularly in the field of human relations, might be useful. Do NOT study agency procedures or detailed manuals. The oral board will be testing your understanding and capacity, not your memory.

5) Get a good night's sleep and watch your general health and mental attitude

You will want a clear head at the interview. Take care of a cold or any other minor ailment, and of course, no hangovers.

What should be done on the day of the interview?

Now comes the day of the interview itself. Give yourself plenty of time to get there. Plan to arrive somewhat ahead of the scheduled time, particularly if your appointment is in the fore part of the day. If a previous candidate fails to appear, the board might be ready for you a bit early. By early afternoon an oral board is almost invariably behind schedule if there are many candidates, and you may have to wait. Take along a book or magazine to read, or your application to review, but leave any extraneous material in the waiting room when you go in for your interview. In any event, relax and compose yourself.

The matter of dress is important. The board is forming impressions about you – from your experience, your manners, your attitude, and your appearance. Give your personal appearance careful attention. Dress your best, but not your flashiest. Choose conservative, appropriate clothing, and be sure it is immaculate. This is a business interview, and your appearance should indicate that you regard it as such. Besides, being well groomed and properly dressed will help boost your confidence.

Sooner or later, someone will call your name and escort you into the interview room. *This is it.* From here on you are on your own. It is too late for any more preparation. But remember, you asked for this opportunity to prove your fitness, and you are here because your request was granted.

What happens when you go in?

The usual sequence of events will be as follows: The clerk (who is often the board stenographer) will introduce you to the chairman of the oral board, who will introduce you to the other members of the board. Acknowledge the introductions before you sit down. Do not be surprised if you find a microphone facing you or a stenotypist sitting by. Oral interviews are usually recorded in the event of an appeal or other review.

Usually the chairman of the board will open the interview by reviewing the highlights of your education and work experience from your application – primarily for the benefit of the other members of the board, as well as to get the material into the record. Do not interrupt or comment unless there is an error or significant misinterpretation; if that is the case, do not hesitate. But do not quibble about insignificant matters. Also, he will usually ask you some question about your education, experience or your present job – partly to get you to start talking and to establish the interviewing "rapport." He may start the actual questioning, or turn it over to one of the other members. Frequently, each member undertakes the questioning on a particular area, one in which he is perhaps most competent, so you can expect each member to participate in the examination. Because time is limited, you may also expect some rather abrupt switches in the direction the questioning takes, so do not be upset by it. Normally, a board member will not pursue a single line of questioning unless he discovers a particular strength or weakness.

After each member has participated, the chairman will usually ask whether any member has any further questions, then will ask you if you have anything you wish to add. Unless you are expecting this question, it may floor you. Worse, it may start you off on an extended, extemporaneous speech. The board is not usually seeking more information. The question is principally to offer you a last opportunity to present further qualifications or to indicate that you have nothing to add. So, if you feel that a significant qualification or characteristic has been overlooked, it is proper to point it out in a sentence or so. Do not compliment the board on the thoroughness of their examination – they have been sketchy, and you know it. If you wish, merely say, "No thank you, I have nothing further to add." This is a point where you can "talk yourself out" of a good impression or fail to present an important bit of information. Remember, *you close the interview yourself.*

The chairman will then say, "That is all, Mr. _______, thank you." Do not be startled; the interview is over, and quicker than you think. Thank him, gather your belongings and take your leave. Save your sigh of relief for the other side of the door.

How to put your best foot forward

Throughout this entire process, you may feel that the board individually and collectively is trying to pierce your defenses, seek out your hidden weaknesses and embarrass and confuse you. Actually, this is not true. They are obliged to make an appraisal of your qualifications for the job you are seeking, and they want to see you in your best light. Remember, they must interview all candidates and a non-cooperative candidate may become a failure in spite of their best efforts to bring out his qualifications. Here are 15 suggestions that will help you:

1) Be natural – Keep your attitude confident, not cocky

If you are not confident that you can do the job, do not expect the board to be. Do not apologize for your weaknesses, try to bring out your strong points. The board is interested in a positive, not negative, presentation. Cockiness will antagonize any board member and make him wonder if you are covering up a weakness by a false show of strength.

2) Get comfortable, but don't lounge or sprawl

Sit erectly but not stiffly. A careless posture may lead the board to conclude that you are careless in other things, or at least that you are not impressed by the importance of the occasion. Either conclusion is natural, even if incorrect. Do not fuss with your clothing, a pencil or an ashtray. Your hands may occasionally be useful to emphasize a point; do not let them become a point of distraction.

3) Do not wisecrack or make small talk

This is a serious situation, and your attitude should show that you consider it as such. Further, the time of the board is limited – they do not want to waste it, and neither should you.

4) Do not exaggerate your experience or abilities

In the first place, from information in the application or other interviews and sources, the board may know more about you than you think. Secondly, you probably will not get away with it. An experienced board is rather adept at spotting such a situation, so do not take the chance.

5) If you know a board member, do not make a point of it, yet do not hide it

Certainly you are not fooling him, and probably not the other members of the board. Do not try to take advantage of your acquaintanceship – it will probably do you little good.

6) Do not dominate the interview

Let the board do that. They will give you the clues – do not assume that you have to do all the talking. Realize that the board has a number of questions to ask you, and do not try to take up all the interview time by showing off your extensive knowledge of the answer to the first one.

7) Be attentive

You only have 20 minutes or so, and you should keep your attention at its sharpest throughout. When a member is addressing a problem or question to you, give him your undivided attention. Address your reply principally to him, but do not exclude the other board members.

8) Do not interrupt

A board member may be stating a problem for you to analyze. He will ask you a question when the time comes. Let him state the problem, and wait for the question.

9) Make sure you understand the question

Do not try to answer until you are sure what the question is. If it is not clear, restate it in your own words or ask the board member to clarify it for you. However, do not haggle about minor elements.

10) Reply promptly but not hastily

A common entry on oral board rating sheets is "candidate responded readily," or "candidate hesitated in replies." Respond as promptly and quickly as you can, but do not jump to a hasty, ill-considered answer.

11) Do not be peremptory in your answers

A brief answer is proper – but do not fire your answer back. That is a losing game from your point of view. The board member can probably ask questions much faster than you can answer them.

12) Do not try to create the answer you think the board member wants

He is interested in what kind of mind you have and how it works – not in playing games. Furthermore, he can usually spot this practice and will actually grade you down on it.

13) Do not switch sides in your reply merely to agree with a board member

Frequently, a member will take a contrary position merely to draw you out and to see if you are willing and able to defend your point of view. Do not start a debate, yet do not surrender a good position. If a position is worth taking, it is worth defending.

14) Do not be afraid to admit an error in judgment if you are shown to be wrong

The board knows that you are forced to reply without any opportunity for careful consideration. Your answer may be demonstrably wrong. If so, admit it and get on with the interview.

15) Do not dwell at length on your present job

The opening question may relate to your present assignment. Answer the question but do not go into an extended discussion. You are being examined for a *new* job, not your present one. As a matter of fact, try to phrase ALL your answers in terms of the job for which you are being examined.

Basis of Rating

Probably you will forget most of these "do's" and "don'ts" when you walk into the oral interview room. Even remembering them all will not ensure you a passing grade. Perhaps you did not have the qualifications in the first place. But remembering them will help you to put your best foot forward, without treading on the toes of the board members.

Rumor and popular opinion to the contrary notwithstanding, an oral board wants you to make the best appearance possible. They know you are under pressure – but they also want to see how you respond to it as a guide to what your reaction would be under the pressures of the job you seek. They will be influenced by the degree of poise you display, the personal traits you show and the manner in which you respond.

ABOUT THIS BOOK

This book contains tests divided into Examination Sections. Go through each test, answering every question in the margin. We have also attached a sample answer sheet at the back of the book that can be removed and used. At the end of each test look at the answer key and check your answers. On the ones you got wrong, look at the right answer choice and learn. Do not fill in the answers first. Do not memorize the questions and answers, but understand the answer and principles involved. On your test, the questions will likely be different from the samples. Questions are changed and new ones added. If you understand these past questions you should have success with any changes that arise. Tests may consist of several types of questions. We have additional books on each subject should more study be advisable or necessary for you. Finally, the more you study, the better prepared you will be. This book is intended to be the last thing you study before you walk into the examination room. Prior study of relevant texts is also recommended. NLC publishes some of these in our Fundamental Series. Knowledge and good sense are important factors in passing your exam. Good luck also helps. So now study this Passbook, absorb the material contained within and take that knowledge into the examination. Then do your best to pass that exam.

EXAMINATION SECTION

EXAMINATION SECTION
TEST 1

DIRECTIONS: Each question or incomplete statement is followed by several suggested answers or completions. Select the one that BEST answers the question or completes the statement. *PRINT THE LETTER OF THE CORRECT ANSWER IN THE SPACE AT THE RIGHT.*

1. In the future there will be a push for all transportation infrastructure to achieve *sustainable development*. An example of *sustainable development* is 1.____

   A. increase the number of lanes on a highway to accommodate added traffic
   B. decrease the number of deaths per million miles of automobile traffic
   C. recycle components of the infrastructure to minimize use of nonrenewable resources
   D. reduce the cost of constructing a mile of highway

2. Pressures are mounting to adopt planning and the principles of TQM. TQM is the abbreviation for 2.____

   A. Transportation Quality Management
   B. Transportation Quality Maintenance
   C. Total Quality Management
   D. Total Quality Maintenance

3. Ladybird Johnson's contribution to highways was 3.____

   A. the planting of wildflowers adjacent to the highway
   B. the planting of trees along the highways
   C. beautifying highway exits with trees and flowers
   D. improving safety at highway exits

4. Bituminous materials used for highways include asphalts and tars derived from destructive distillation of materials such as coal and wood. Tars have been little used recently primarily because of 4.____

   A. high cost of the tar
   B. inability to meet the specifications for tar
   C. difficulty in applying this material
   D. lack of availability of the material

5. Contraction joints in a Portland cement concrete highway slab are provided in order to 5.____

   A. allow the slab to crack at the joint
   B. minimize hydroplaning in wet weather
   C. absorb expansion of the slab
   D. control alligator cracks on the surface of the slab

6. Of the following, the one that is designated as a grade of asphalt binder is 6.____

   A. AD B. AS C. PR D. PG

7. A property of asphaltic materials is viscosity. Viscosity in a liquid is the 7.____

A. resistance to evaporation
B. tendency to separate into its components
C. ability of the liquid to mix with other materials
D. tendency of the liquid to resist flow

8. Medium-curing cutback asphalt contains a _______-type solvent. 8.____

A. kerosene
B. naphtha
C. heavy fuel oil
D. benzene

9. Asphalt emulsions are becoming the preferred asphalt binder in many agencies because 9.____

A. emulsions are easier to apply
B. of concern about hydrocarbon emissions from cut-back asphalts
C. emulsions produce an asphalt concrete that is more resistant to abrasion
D. emulsions are more resistant to water penetration

10. An example of vector control on a roadside is 10.____

A. eliminating a breeding ground for rodent populations
B. keeping signs and directions in good condition
C. keeping the roadside free of litter and debris
D. halting erosion on the roadside

11. An attenuator in highway work is a(n) 11.____

A. warning device during highway maintenance work
B. barrier to protect highway maintenance workers
C. overhead warning to slow down automobiles
D. crash safety board

12. Preventive maintenance is a planned strategy of cost effective treatment that preserves the system.
It can BEST be expressed as 12.____

A. don't fix it if it isn't broken
B. a stitch in time saves nine
C. if something can go wrong it will
D. it is preferable to replace than repair

13. There are two types of aggregate used in asphalt mixes: crushed aggregate and round aggregate. Of the following statements relating to the two types of aggregate, the one that is CORRECT is: 13.____

A. There is no advantage in either aggregate.
B. Round aggregate is preferable to crushed aggregate.
C. Crushed aggregate is preferable to round aggregate.
D. It all depends on the source of the aggregate.

14. Temperature limits should be strictly observed when using asphalt cements. The mixing temperature should be between 14.____

A. 125° F and 175° F
B. 175° F and 225° F
C. 275° F and 325° F
D. 325° F and 375° F

15. When high early strength is desired in Portland cement concrete, the cement to use is type 15.____

A. II
B. III
C. IV
D. V

16. The main purpose of grading aggregates in an asphalt roadway mix is to 16.____

A. provide a good surface
B. provide a strong mix
C. minimize the quantity of asphalt required
D. provide a dense mix to prevent water seepage in the roadway

17. The wearing quality of an aggregate is determined by testing for resistance to 17.____

A. abrasion
B. crushing
C. chemical deterioration
D. frost

18. According to the AASHTO Maintenance Manual on Roadways, earth-aggregate roadway surfaces and subsurfaces are most effective when they have achieved at least _______% of their compaction capacity. 18.____

A. 80
B. 85
C. 90
D. 95

19. Superelevation on a highway usually occurs 19.____

A. on an approach to a bridge
B. on a horizontal highway curve
C. at a high point on a vertical highway curve
D. at a low point on a vertical highway curve

20. Shoulders have two basic purposes in the roadway system: they provide lanes for emergency or safe travel and they 20.____

A. prevent vegetation encroaching on the roadway surface
B. allow seepage into the subgrade of runoff from the roadway surface
C. provide room for barriers
D. provide lateral support to the pavement structure

21. For roadway cross-sections without curbs, shoulder cross slopes usually range from _______ for paved surfaces. 21.____

A. 1% to 3%
B. 3% to 5%
C. 5% to 7%
D. 7% to 9%

22. If turf shoulders are used on a roadway, the advantage of using native grasses usually is that native grasses 22.____

A. do not need mowing
B. are more pleasing to the eye than imported grasses
C. are less-expensive than other grasses
D. do not need irrigation

23. A deep rut in an aggregate shoulder at the edge of a hard-surfaced roadway can usually be corrected by reshaping the shoulder MOST efficiently with a 23.____

A. scarifier
B. bulldozer
C. motor grader
D. roller

24. A method of scheduling shoulder maintenance in response to risk management efforts is to 24.____

A. maximize the life of the shoulder
B. minimize the labor cost of maintaining the shoulder in a serviceable condition
C. defend maintenance policies and practices in the event of accident litigation involving the shoulders
D. use a database to predict when maintenance and repair will be required for the shoulders

25. Spalling of the surface of a concrete roadway is generally caused by 25.____

A. inadequate vibrating when pouring the concrete
B. too low a water/cement ratio in the concrete mix
C. expansion of the concrete
D. the use of epoxy-coated reinforcing bars in the concrete

KEY (CORRECT ANSWERS)

1.	C	11.	D
2.	C	12.	B
3.	A	13.	C
4.	D	14.	C
5.	A	15.	B
6.	D	16.	C
7.	D	17.	A
8.	A	18.	D
9.	B	19.	B
10.	A	20.	D

21. B
22. D
23. C
24. C
25. C

TEST 2

DIRECTIONS: Each question or incomplete statement is followed by several suggested answers or completions. Select the one that BEST answers the question or completes the statement. *PRINT THE LETTER OF THE CORRECT ANSWER IN THE SPACE AT THE RIGHT.*

1. Corrosion of reinforcing bars in a reinforced concrete road pavement can be caused by water entering the concrete road pavement or ______ entering the concrete road pavement. 1.____

   A. sulfates B. chlorides C. carbonates D. fluorides

2. A tack coat applied to an old roadway surface creating a bond between the old and the new surface is applied with a distributor at the rate of ______ gallon per square yard. 2.____

   A. .05 to .15 B. .15 to .25
   C. .25 to .35 D. .35 to .45

3. Some voids should be included in compacted asphaltic concrete to allow for expansion in hot weather. Surface course voids is usually recommended to be 3.____

   A. 2% to 4% B. 4% to 6% C. 6% to 8% D. 8% to 10%

4. The highest temperature that asphalts can withstand is 4.____

   A. 300° F B. 350° F C. 400° F D. 450° F

5. A polymer is a substance containing 5.____

   A. microscopic air bubbles
   B. the element silicon
   C. a definite lattice arrangement
   D. giant molecules

6. The one of the following that is not a polymer concrete is ______ concrete. 6.____

   A. epoxy B. methyl methacrylate
   C. polyurethane D. vermiculite

7. It is suspected that a given stretch of existing roadway may have substructure problems; that is, the soil under the roadway is weak. 7.____
   Of the following, the BEST method of testing the sub-surface is with a ______ test.

   A. falling weight deflectometer
   B. soil porosity
   C. consolidation settlement
   D. tiltmeter

8. The one of the following that is NOT a claimed potential advantage of using reclaimed asphalt paving for a new asphalt pavement is 8.____

   A. energy saving
   B. cost reduction
   C. stronger asphalt pavement
   D. conservation of natural resources

9. Mineral dust is added to asphalt primarily to ______ asphalt. 9.____

A. make it easier to roll the
B. increase the viscosity of the
C. stabilize the
D. eliminate air pockets in

10. The smallest size sieve that mineral dust should pass through is No. 10.____

A. 50 B. 100 C. 150 D. 200

11. Most asphalt used in highway construction is derived from 11.____

A. natural sources
B. coal distillation
C. inorganic materials
D. petroleum distillation

12. Asphalt emulsions should be cationic or anionic depending on 12.____

A. the pH value of the asphalt
B. aggregate size distribution
C. the pH value of the water used
D. the type of aggregates used

13. Air-entrained cement is used in concrete to 13.____

A. expose concrete to severe frost action
B. give the concrete high early strength
C. resist sulfate deterioration
D. make the concrete more workable

14. The main purpose of vibrating poured concrete is to 14.____

A. eliminate air pockets in the placed concrete
B. prevent segregation in the concrete
C. allow excess water to rise to the surface
D. decrease the water/cement ratio in the poured concrete

15. The primary purpose of curing a concrete road slab shortly after pouring is primarily to 15.____

A. prevent loss of water in the concrete due to evaporation
B. protect the concrete from changes in outside temperature
C. shield the surface of the concrete against the loss of heat
D. prevent segregation in the concrete slab

16. A bag of Portland cement weighs MOST NEARLY ______ pounds. 16.____

A. 82 B. 86 C. 90 D. 94

17. The water/cement ratio for 4,000 pounds per square inch concrete is, in gallons of water per bag of concrete, MOST NEARLY 17.____

A. 3 B. 5 C. 7 D. 9

Questions 18-19.

DIRECTIONS: Questions 18 and 19 refer to the notes shown below.

The notes shown below are used to determine the elevation of the top of a manhole M.

Point	BS	HI	F.S.	Elevation
BMA	0.72			151.42'
Manhole M			4.25	

18. The elevation of manhole M is, in feet and inches, MOST NEARLY 18.____

A. 140.45 B. 148.95 C. 153.89 D. 162.39

19. The elevation of manhole M, in feet and inches, is MOST NEARLY 19.____

A. 148'-5 3/8" B. 148'-11 3/8"
C. 153'-10 11/16" D. 162'-4 11/16"

20. 20.____

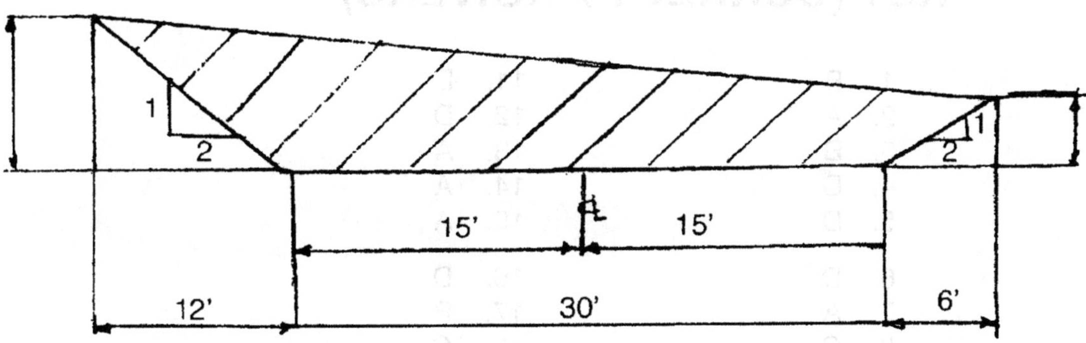

The cross-section area of the excavation shown above is, in square feet, MOST NEARLY

A. 171 B. 175 C. 179 D. 183

21. The cross-section area of the sewer section is, in square feet, MOST NEARLY 21.____

A. 12.2
B. 12.4
C. 12.6
D. 12.8

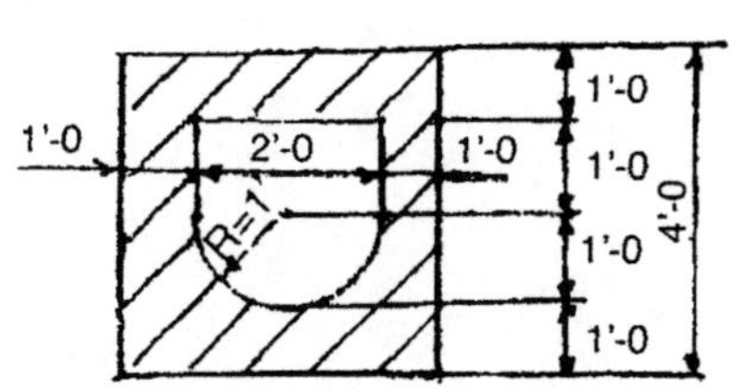

22. 3.66 meters is MOST NEARLY ______ feet. 22.____

A. 11 B. 12 C. 13 D. 14

23. In some states litigation has established a legal definition of a safety hazard as any pavement dropoff exceeding 5.1 centimeters. This, in inches, is MOST NEARLY 23.____

A. 1 B. $1\frac{1}{2}$ C. 2 D. $2\frac{1}{2}$

24. In a roadway maintenance manual is a subject heading titled AESTHETIC OBJECTIVES. An example of an aesthetic objective is to 24.____

A. reduce highway accidents
B. enhance a roadway's scenic qualities
C. improve the roadway's drainage system
D. widen the roadway at a turnout to prevent vehicles backing up into the highway

25. A cubic yard of ordinary unreinforced concrete weighs MOST NEARLY _____ pounds. 25.____

A. 2000 B. 3000 C. 4000 D. 5000

KEY (CORRECT ANSWERS)

1.	B	11.	D
2.	A	12.	D
3.	B	13.	A
4.	D	14.	A
5.	D	15.	A
6.	D	16.	D
7.	A	17.	B
8.	C	18.	C
9.	C	19.	C
10.	D	20.	A

21. B
22. B
23. C
24. B
25. C

EXAMINATION SECTION
TEST 1

DIRECTIONS: Each question or incomplete statement is followed by several suggested answers or completions. Select the one that BEST answers the question or completes the statement. *PRINT THE LETTER OF THE CORRECT ANSWER IN THE SPACE AT THE RIGHT.*

1. In a hand tap set, the tap used to start a thread in a drilled hole is known as a ______ tap. 1.____

   A. taper B. plug C. small D. bottoming

2. The type of fastener used to fasten thin gauge metal to wood backing, without drilling, is known as a 2.____

   A. sheet metal screw
   B. cap screw
   C. wood screw
   D. screw nail

3. Set screws are usually used for 3.____

   A. fastening collars to shafts
   B. holding thin metal sheets together
   C. holding roller bearings to shafts
   D. clamping together steel angles

4. When using a pedestal type grinding wheel, the operator should always 4.____

   A. have the work-rest loose
   B. avoid striking the rotating wheel
   C. increase the speed above normal
   D. use a respirator

5. A flat cold chisel is the type of chisel usually used for chipping and/or cutting 5.____

   A. filleted corners
   B. flat surfaces
   C. V-shaped grooves
   D. narrow grooves

6. The type of chisel that is usually used to cut keyways in cast iron is generally known as a ______ chisel. 6.____

   A. star
   B. cold
   C. cape
   D. diamond point

7. The primary difference between brazing and soldering is that brazing requires 7.____

   A. greater heat
   B. a smaller soldering iron
   C. the use of soft solder
   D. the use of 50-50 solder

Questions 8-9.

DIRECTIONS: Questions 8 and 9 refer to the sketch below.

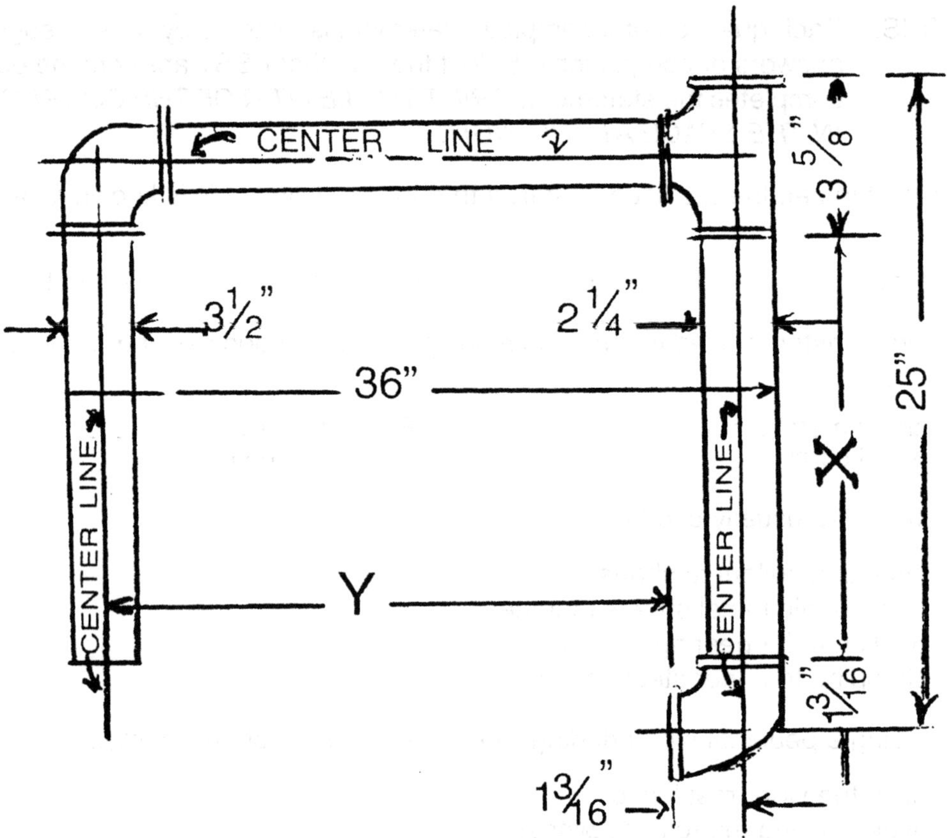

8. In the above sketch, the dimension X, in inches, is 8.___

   A. 19 13/16 B. 20 3/16 C. 20 3/8 D. 21 3/16

9. In the above sketch, the dimension Y, in inches, is 9.___

   A. 30 7/16 B. 31 5/16 C. 31 7/16 D. 31 15/16

10. A piece of 4" cast iron pipe may BEST be cut with a 10.___

    A. hacksaw having a blade with 32 teeth per inch
    B. hacksaw having a blade with 14 teeth per inch
    C. hammer and round nose chisel
    D. hammer and diamond point chisel

11. A four-inch length of straight pipe, threaded on both ends, is generally called a 11.___

    A. sleeve B. nipple C. stud D. extension

12. A pipe fitting that is usually used to join together two threaded pipes of the same diameter is known as a 12.___

    A. union
    B. straight T with reducer leg
    C. nipple
    D. straight tee

13. A pipe fitting that is generally used to join two threaded pipes of different diameters is called a(n) 13.____

A. close nipple
B. union
C. adapter
D. reducer

14. A 90° pipe fitting that has a male thread at one end and a female thread at the other end is generally known as a 90° 14.____

A. elbow
B. street elbow
C. reducing ell
D. long radius ell

15. Paints generally used for covering outside pipes or sheet iron are composed of 15.____

A. mineral pigments, organic vehicles and thinners
B. resins dissolved in organic thinners
C. pigmented oil and linseed oil
D. lac gum dissolved in alcohol

16. A painted panel of wood after being exposed to the atmosphere becomes leather-like in appearance. 16.____
This paint failure is called

A. checking
B. alligatoring
C. wrinkling
D. chalking

17. Paint brushes that are used for alkyd paints are usually cleaned with 17.____

A. soap and water
B. turpentine and mineral spirits
C. linseed oil
D. denatured alcohol mixed with water

18. Of the following statements, the one which is INCORRECT concerning painting practices is: 18.____

A. Zinc dust primers are used for galvanized iron and sheet zinc
B. Red lead paint is usually used as a final coat for steel surfaces
C. Rubber-base paints may be applied to dry or damp walls
D. Freshly varnished work should be kept clean and in a dust-free space

19. A white paint, that can cover 500 square feet of surface per gallon, is used to paint the crosswalks at street intersections. 19.____
If the area at each intersection is equal to 300 square feet, the number of gallons required to paint 50 intersections is MOST NEARLY

A. 10
B. 20
C. 30
D. 40

20. Of the following methods of splicing insulated electrical wires, the one which is recommended is to strip the ends, twist them together, 20.____

A. and cover with friction tape
B. solder and cover with friction tape
C. shellac and cover with rubber tape
D. solder, cover with rubber tape, and then with friction tape

21. If two 120V incandescent lamps are connected in parallel in a 120V circuit, the result will MOST likely be that the 21.____

A. lamps will light up to normal brilliancy
B. voltage across each lamp will be reduced to 60 volts
C. life of each lamp will be doubled
D. lamp will light up to 1/2 their normal brilliancy

22. Portable electric hand tools are usually polarized by means of a(n) 22.____

A. circuit breaker
B. fuse
C. three-prong plug
D. overload switch

23. The flux generally used when soldering electrical copper connections is 23.____

A. zinc chloride
B. an alcoholic solution of resin
C. muriatic acid
D. stearin

24. Fuses in the electric wiring systems of a car or truck are MAINLY used for the purpose of 24.____

A. making it easy to disconnect some of the lights while allowing others to burn
B. reducing the amount of current used, in order to save the battery
C. automatically opening the circuit in case of an overload
D. preventing the battery from overcharging under high speed

25. The BEST way to fasten electric conduit to an outlet box is by means of a 25.____

A. bushing on the end of the conduit
B. locknut on the outside of the box
C. bushing on the inside and a locknut on the outside
D. locknut on the inside and a bushing on the outside

26. The GREATEST hazard of explosion exists whenever 26.____

A. gasoline is stored in airtight tanks
B. a pool of gasoline is exposed to air
C. gasoline is in a partially-full closed tank
D. gasoline comes into contact with oil

27. 27.____

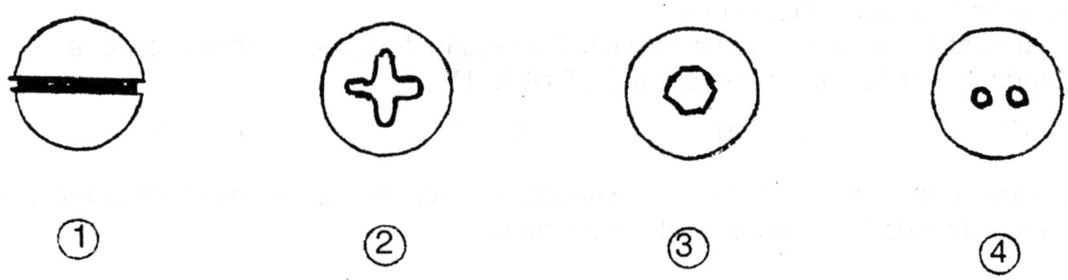

In the above sketch, the head of a screw which represents an alien-head screw is numbered

A. 1
B. 2
C. 3
D. 4

28. A ratchet wrench is usually used when 28.____

A. the surface finish of a bolt must be preserved
B. only a short swing of the wrench handle is permissible
C. nuts are practically inaccessible
D. tightening compression fittings

29. Grout in construction work is usually used to 29.____

A. increase the strength of concrete
B. seal porous timber surfaces
C. prime concrete sidewalks
D. fill spaces between brick or stone joints

30. A cutting tool that is being ground on an emery wheel is usually cooled by immersing it in 30.____

A. oil
B. water
C. kerosene
D. turpentine

31. Of the following statements concerning the use of screwdrivers, the one which is INCORRECT is: 31.____

A. Always use a screwdriver with a blade that fits the screw to be turned
B. Hold the work in one hand while turning the screwdriver with the other
C. A screwdriver with an insulated handle should be used for making electrical repairs
D. A screwdriver should not be used as a chisel or hammer

32. A miter box is usually used for 32.____

A. making diagonal cuts
B. holding the flux in soldering
C. storing small machine screws
D. storing precision tools

Questions 33-38.

DIRECTIONS: Questions 33 through 38, inclusive, should be answered in accordance with the following paragraph.

It is important that traffic signals be regularly and effectively maintained. Signals with impaired efficiency cannot be expected to command desired respect. Poorly maintained traffic signs create disrespect in the minds of those who are to obey them and thereby reduce the effectiveness and authority of the signs. Maintenance should receive paramount consideration in the design and purchase of traffic signal equipment. The initial step in a good maintenance program for traffic signals is the establishment of a maintenance record. This record should show the cost of operation and maintenance of different types of equipment. It should give complete information regarding signal operations and indicate where defective planning exists in maintenance programs.

33. The word *effectively*, as used in the above paragraph means MOST NEARLY 33.____

A. occasionally
B. properly
C. expensively
D. cheaply

34. The word *impaired,* as used in the above paragraph, means MOST NEARLY 34.____

A. reduced B. increased C. constant D. high

35. The word *desired,* as used in the above paragraph, means MOST NEARLY 35.____

A. public B. complete C. wanted D. enough

36. The word *paramount,* as used in the above paragraph, means MOST NEARLY 36.____

A. little B. chief C. excessive D. some

37. The word *initial,* as used in the above paragraph, means MOST NEARLY 37.____

A. first
B. final
C. determining
D. most important

38. The word *defective,* as used in the above paragraph, means MOST NEARLY 38.____

A. suitable B. real C. good D. faulty

39. A half round file is usually used for 39.____

A. removing stock rapidly
B. clearing out square corners
C. finishing the bottoms of narrow slots
D. finishing concave surfaces

40. For finishing flat metal surfaces, the type of file usually used is the ______ file. 40.____

A. pillar B. hand C. square D. drill

KEY (CORRECT ANSWERS)

1.	A	11.	B	21.	A	31.	B
2.	D	12.	A	22.	C	32.	A
3.	A	13.	D	23.	B	33.	B
4.	B	14.	B	24.	C	34.	A
5.	B	15.	A	25.	C	35.	C
6.	C	16.	C	26.	B	36.	B
7.	A	17.	B	27.	C	37.	A
8.	B	18.	B	28.	B	38.	D
9.	D	19.	C	29.	D	39.	D
10.	D	20.	D	30.	B	40.	B

TEST 2

DIRECTIONS: Each question or incomplete statement is followed by several suggested answers or completions. Select the one that BEST answers the question or completes the statement. *PRINT THE LETTER OF THE CORRECT ANSWER IN THE SPACE AT THE RIGHT.*

1. The type of pliers usually used for holding or bending thin flat iron stock is known as 1.____

   A. diagonal pliers B. round nose pliers
   C. nippers D. side cutting pliers

2. An auger bit is usually used for boring a hole in 2.____

   A. brass B. concrete C. wood D. steel

3. The dimension 45" expressed in feet is MOST NEARLY 3.____

   A. 3 1/3 B. 3 1/2 C. 3 3/4 D. 3 7/8

4. The type of hand saw that is used to cut wood along the grain is generally known as a ______ saw. 4.____

   A. band B. rip C. back D. cross cut

5. The LEAST likely cause for the breaking of hacksaw blades is 5.____

   A. using a fine-toothed blade on thin work
   B. using a coarse-toothed blade on thin work
   C. working with a blade that is tightly drawn in the hacksaw frame
   D. applying too much pressure on the work

6. A 1:3:5 mixture of concrete generally refers to a mixture of 1 part of ______, 3 parts of ______, 5 parts of ______. 6.____

   A. gravel, sand, cement B. sand, cement, gravel
   C. water, cement, gravel D. cement, sand, gravel

7. 85 percent of $5,250 is MOST NEARLY 7.____

   A. $3,463.50 B. $4,361.50 C. $4,462.50 D. $4,666.50

8. A scriber is usually used for 8.____

   A. starting a hole in iron B. measuring lengths
   C. cleaning threads D. layout work

9. The gauge of an iron sheet indicates its 9.____

   A. thickness B. length
   C. weight per square inch D. width

10. The type of gears used on the ends of two intersecting shafts 90° to each other for transmitting motion are known as 10.____

    A. spur B. bevel C. spline D. spiral

11. If the shortest distance between the edges of two holes drilled in a flat steel plate is I 1/2" and the diameters of the holes are 3/4" and 1", the distance between centers is MOST NEARLY 11.____

A. 2 1/8" B. 2 1/4" C. 2 3/8" D. 2 3/4"

12. A flat head screw that is identified as a 1/4X 20 - 1" long screw is MOST likely a ______ screw. 12.____

A. wood
B. sheet metal
C. cap
D. machine

13. If your foreman informs you that a traffic signal was obliterated, he MOST likely means that the traffic signal was 13.____

A. stolen B. obsolete C. loose D. destroyed

14. Of the following statements, the one which is MOST correct concerning a *regulatory* traffic sign is that the sign 14.____

A. if disregarded by the driver, is punishable as a misdemeanor
B. calls attention to conditions that are potentially hazardous to traffic
C. shows route designations and directions
D. shows points of interest and other geographical information

15. The MAIN reason for the alternate-side-of-the-street parking regulations is to 15.____

A. facilitate the cleaning of streets
B. allow room for moving traffic
C. allow room for delivery trucks
D. provide space for children to play

16. The geometrical shape of *STOP* signs if 16.____

A. octagonal
B. triangular
C. diamond
D. rectangular

17. In an engine, the MAIN purpose for using oil as a lubricant is to keep 17.____

A. the engine parts from rusting
B. a film between the moving parts
C. the internal parts clean, by flushing them
D. the vibration down

18. The QUICKEST method of determining a defective spark plug is to 18.____

A. take out the spark plugs and examine them
B. drive to the garage and let a mechanic tell you
C. short circuit the spark plugs one at a time with an insulated screwdriver
D. replace all the spark plugs with new ones

19. The device that controls the charging rate of a generator in the generator-battery circuit of an automotive engine is usually the 19.____

A. ignition coil
B. generator regulator
C. condenser
D. generator solenoid

20. The device used with a gasoline engine to change the liquid fuel into vapor and mix it with air is called a(n) 20.____

A. fuel pump
B. automatic choke
C. carburetor
D. vapor regulator

21. The gap between the electrodes of a spark plug is usually measured with a 21.____

A. feeler gauge
B. flat stock
C. depth gauge
D. caliper

22. The proper spark plug gap for MOST truck engines is approximately 22.____

A. .015" B. .018" C. .030" D. .042"

23. A truck mounted air compressor that supplies air to a number of pneumatic tools is usually set to deliver air at APPROXIMATELY ______ psi. 23.____

A. 30 B. 50 C. 90 D. 160

Questions 24-25.

DIRECTIONS: Questions 24 and 25 refer to the sketch below depicting a street intersection.

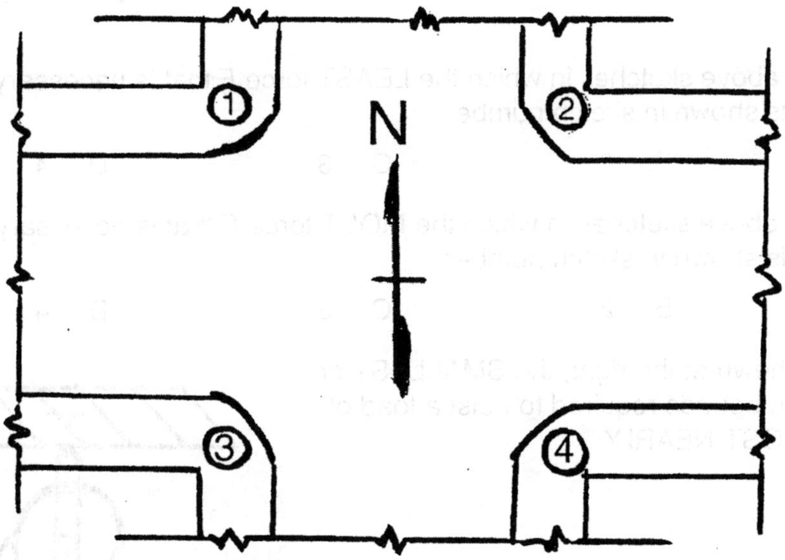

24. In the above sketch, the southeast corner is numbered 24.____

A. 1 B. 2 C. 3 D. 4

25. In the above sketch, the northwest corner is numbered 25.____

A. 1 B. 2 C. 3 D. 4

26. The proper position to place yourself when lifting a heavy box from the floor is to 26.____

A. squat down, bend knees, keep back straight and lift
B. bend down, hunch back, and lift

C. keep feet away from object, bend back, and lift
D. bend, and use the back muscles when lifting

Questions 27-28.

DIRECTIONS: Questions 27 and 28 refer to the sketches immediately below.

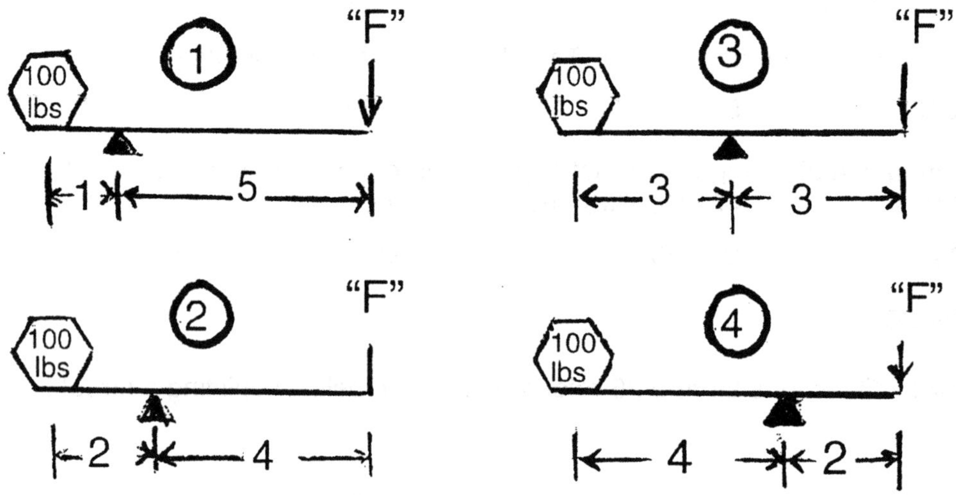

27. The one of the above sketches in which the LEAST force *F* that is necessary to raise the 100 lb. weight is shown in sketch number 27._____

A. 1 B. 2 C. 3 D. 4

28. The one of the above sketches in which the MOST force *F* that is necessary to raise the 100 lb. weight is shown in sketch number 28._____

A. 1 B. 2 C. 3 D. 4

29. In the sketch shown at the right, the SMALLEST or LEAST pull *P* in pounds required to hoist a load of 1 1/2 tons is MOST NEARLY 29._____

A. 500
B. 1,000
C. 1,500
D. 2,000

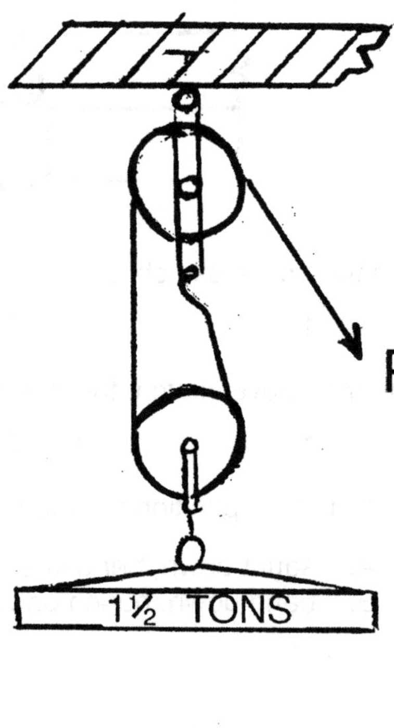

30. Of the following statements, the one which is INCORRECT is: Fibre rope 30.____

A. should be stored in an air-tight container
B. should never be stored on the ground
C. is approximately 10% stronger when wet
D. should be inspected periodically

31. The term *mousing,* as used in rigging, means MOST NEARLY 31.____

A. temporarily attaching a rope to a hook
B. attaching a rope to a tackle block
C. securing a grip on a rope under strain
D. placing rope yarn on a hook to prevent load from becoming detached

32. The safe distance that the bottom of a ladder, the top of which is placed against a wall or pole must extend out from the base of the wall or pole, is usually ______ the length of the ladder. 32.____

A. 1/2 B. 1/4 C. 1/6 D. 1/8

33. Of the following painting materials, the one which should NOT be used to treat or cover wood ladders is 33.____

A. linseed oil
B. oil paint
C. shellac
D. varnish

34. A knot that has a non-slipping eye, will not jam, and is easily untied is generally known as a(n) 34.____

A. sheet bend
B. granny
C. bowline
D. overhand knot

35. A sheepshank knot is PRINCIPALLY used for 35.____

A. shortening a rope without cutting it
B. fastening a rope at right angles to a post
C. attaching a rope to a ring
D. joining large hawsers

36. Of the following procedures for softening a piece of steel, good practice is to FIRST heat the steel and then 36.____

A. cool it rapidly
B. dip it in cold water
C. cool it slowly
D. cool it in a cold solution of salt

37. Of the following types of measuring rules, the one which can BEST be used to measure directly the circumference of a 4-inch diameter pipe is the ______ rule. 37.____

A. zig-zag
B. folding
C. caliper
D. push-pull

38. The painting of traffic safety lines and pedestrian crosswalks at busy intersections is one of the jobs of a traffic device maintainer. The work crew usually protects itself from traffic by 38.____

A. re-routing the traffic at the next intersection
B. lengthening the time for the *Stop* signal on the traffic light
C. wearing bright yellow work clothes
D. placing safety cones to divert traffic away from the work area

39. An unloader is a device that is usually found on a(n) 39.____

A. pneumatic tool
B. air compressor
C. storage tank
D. block and fall

40. A device that can be used repeatedly for marking out shapes on materials is generally called a 40.____

A. blueprint
B. tracing
C. scantling
D. template

KEY (CORRECT ANSWERS)

1.	D	11.	C	21.	A	31.	D
2.	C	12.	D	22.	C	32.	B
3.	C	13.	D	23.	C	33.	B
4.	B	14.	A	24.	D	34.	C
5.	A	15.	A	25.	A	35.	A
6.	D	16.	A	26.	A	36.	C
7.	C	17.	B	27.	A	37.	D
8.	D	18.	C	28.	D	38.	D
9.	A	19.	B	29.	C	39.	B
10.	B	20.	C	30.	A	40.	D

EXAMINATION SECTION
TEST 1

DIRECTIONS: Each question or incomplete statement is followed by several suggested answers or completions. Select the one that BEST answers the question or completes the statement. *PRINT THE LETTER OF THE CORRECT ANSWER IN THE SPACE AT THE RIGHT.*

1. When making a preliminary inspection of a new street marking job, the FIRST thing to check is whether 1.____

   A. the location is correct
   B. all dimensions are correct
   C. the right paint is specified
   D. traffic can easily be controlled

2. After a preliminary inspection of a new street marking job has been made and it has been found that it can be laid out exactly as shown in the drawings received from Plans and Surveys, the site should be reinspected on the first day of actual work to check that 2.____

   A. the dimensions are correct according to the plans
   B. the orientation has not changed
   C. excavation work that did not exist on his first inspection does not obstruct his work
   D. the traffic can easily be controlled

3. Of the following, it is MOST important when inspecting the installation of a sign in a garage or on a street to check for the ______ the sign. 3.____

   A. correct width of
   B. correct area of
   C. correct mounting height of
   D. removal of all scuff marks below

4. When inspecting a job site in an off-street parking garage prior to starting a new job involving markings, the FIRST thing to look for is 4.____

   A. obstructions such as beams which will require that the layout be altered
   B. oil on the floor
   C. paint splashes on the floor
   D. vehicles which must be moved

5. The one of the following items which should be checked on a job involving the installation of custom-made highway guide signs but which need NOT be checked during the installation of street regulatory signs is the ______ of the sign. 5.____

   A. color
   B. wording and spelling
   C. width
   D. area

6. Assume that you are facing east while standing on the northwest corner of the intersection of two streets. One of these streets runs north and south, and the other runs east and west. 6.____
   The SOUTHWEST corner of this intersection is

A. *directly* across the street in front of you
B. *directly* across the street to your right
C. *diagonally* across the intersection from you
D. *directly* across the street to your left

7. A street running north and south intersects a street running east and west. Four men designated as A, B, C, and D are each on a different corner of the intersection. A is on the NW corner and faces east; B is on the SW corner and faces north; C is on the SE corner and faces west; and D is on the NE corner and faces west. The two men who are facing DIRECTLY toward each other are 7.____

A. A and B B. B and C C. C and D D. A and D

8. Of the following, the MOST important item to check during a routine inspection of an air compressor is the 8.____

A. amount of air used daily
B. number of hours it has been operated
C. diaphragm diameter
D. condition of the paint finish

9. Assume that a crew assigned to you goes out to paint some street markings on a street which has a great deal of traffic. The traffic should be diverted away from the working area by means of 9.____

A. Class I barricades
B. Class II barricades
C. Class I barricades and cones
D. cones

10. Assume that an extensive area within an off-street parking facility has caved in. Until repairs are completed, cars should be kept away from this area by means of 10.____

A. Class I barricades
B. Class I barricades and flasher lights
C. Class II barricades and cones
D. warning signs and Class I barricades

11. A line of traffic cones, being used to divert traffic fron men painting cross-walks in the lane nearest the curb, should begin at the curb at a point whose distance fron the working area is _______ feet, and the cones should be _______ feet apart. 11.____

A. 40; 10 B. 60; 15 C. 80; 15 D. 100; 10

12. Crews doing street marking work at night should wear 12.____

A. gray coveralls and set out traffic cones to divert traffic away from the area
B. reflectorized vests and set out traffic cones to divert traffic away from the area
C. bright yellow helmets and gray coveralls
D. bright blue helmets and set out traffic cones to divert traffic away from the area

13. Assume that the top of a 12 foot ladder is to be placed against a wall. The RECOMMENDED safe practice is that the ladder should be placed so that the distance from the bottom of the ladder to the base of the wall is ______ ft. 13.____

A. 1 B. 2 C. 3 D. 5

14. According to the State Vehicle and Traffic Law, when driving at a speed of 40 miles per hour along a dry road, the driver should maintain a distance between his car and the car immediately ahead of him of AT LEAST ______ car lengths. 14.____

A. 2 B. 3 C. 4 D. 5

15. Assume that a man has been knocked unconscious. Which of the following should NOT be done to the victim? 15.____

A. Give him something to drink
B. Hold a handkerchief with spirits of ammonia under his nose if he is breathing
C. Keep him covered with a blanket
D. Give him artificial respiration if he is not breathing

16. A paint sprayer may have gauges showing the pressure of the tank, the paint pressure, and the atomizer pressure. When the sprayer is operating properly, the 16.____

A. paint pressure is higher than the tank pressure
B. atomizer pressure is higher than the tank pressure
C. paint and atomizer pressures are equal
D. atomizer pressure is higher than the paint pressure

17. A certain paint can cover 310 square feet per gallon. The number of gallons of this paint required to paint 200 lines each 6 inches wide and 18 feet-6 inches long is MOST NEARLY 17.____

A. 2 B. 4 C. 6 D. 8

18. Paint brushes that are used with an oil-based paint are USUALLY cleaned with 18.____

A. turpentine
B. linseed oil
C. acetone
D. alcohol

19. The air in an air compressor cylinder is DIRECTLY compressed by the 19.____

A. pressure switch
B. surge chamber
C. cam
D. piston

20. The part which permits the motor of an air compressor to start free of load regardless of the tank pressure is the 20.____

A. unloader valve
B. surge tank
C. pressure switch
D. drain cock

21. Assume that instead of spraying paint properly, a paint sprayer ejects a solid stream of paint from its nozzle. The one of the following that may cause this condition is 21.____

A. compressor tank pressure is too high
B. compressor tank pressure is lower than the atomizer pressure
C. atomizer pressure is higher than the paint pressure
D. atomizer pressure is too low

22. The one of the following which is a *regulatory* sign is the 22.____

A. bump sign
B. low clearance sign
C. route marker
D. stop sign

23. The one of the following which is a *regulatory* sign is the ______ sign. 23.____

A. yield
B. stop ahead
C. side road
D. slippery when wet

24. The one of the following signs which is octagonal is the ______ sign. 24.____

A. speed limit
B. stop ahead
C. road narrows
D. stop

25. Of the following statements, the one which gives the function of a *warning* sign is that this sign 25.____

A. indicates route designations, destinations, or distances
B. gives the driver notice of laws or regulations that apply at a given place, disregard of which is punishable as a violation or a misdemeanor
C. calls attention to conditions in or adjacent to a street that are potentially hazardous , to traffic
D. indicates points of interest or geographical locations

26. The regulation manual on temporary traffic control of the department of traffic defines Class II barricades as being of the *horse* type with only one rail.
It further specifies that the rail should be marked on 26.____

A. *one* side with 3" vertical red and white, black and white, or black and yellow reflectorized stripes
B. *both* sides with 3" vertical red and white, black and white, or black and yellow stripes
C. *both* sides with 6" reflectorized red and white, black and white, or black and yellow stripes sloping at an angle of 45
D. *both* sides with 6" vertical red and white or black and white stripes

27. Silk screening is a method of 27.____

A. temporarily concealing signs already erected but not ready to be used
B. painting signs
C. protecting newly painted crosswalks until they dry
D. protecting reflectorized signs from dust

28. The blade of a snow plow is USUALLY made of 28.____

A. monel
B. steel
C. tungsten carbide
D. beryllium

29. To PROPERLY check the lifting device of a snow plow at the beginning of the snow season, the plow blade should be 29.____

A. raised and kept in that position for at least three minutes in order to detect leaks in the system
B. raised by the lifting device once to see if it operates

C. dropped quickly after being brought to the raised position
D. raised and lowered and then the operation should be repeated

30. At the present time, the department of traffic USUALLY reflectorizes signs by 30.____

A. coating the portion of the sign to be reflectorized with very tiny glass beads held by an adhesive base
B. outlining the reflectorized portion of the sign with large glass *bull's eyes*
C. making the reflectorized portion of the sign with *Scotch Lite*
D. painting the reflectorized portion of the sign with *Luminar*

KEY (CORRECT ANSWERS)

1. A	11. D	21. D
2. C	12. B	22. D
3. C	13. C	23. A
4. A	14. C	24. D
5. B	15. A	25. C
6. B	16. D	26. C
7. D	17. C	27. B
8. B	18. A	28. B
9. D	19. D	29. A
10. C	20. A	30. C

TEST 2

DIRECTIONS: Each question or incomplete statement is followed by several suggested answers or completions. Select the one that BEST answers the question or completes the statement. *PRINT THE LETTER OF THE CORRECT ANSWER IN THE SPACE AT THE RIGHT.*

1. The material which causes the hydraulic plunger of a heavy duty hydraulic jack to move is 1.____
   A. oil b. petrolatum C. alcohol D. glycerol

2. "Vapor Lock" will DIRECTLY affect the operation of 2.____
   A. air compressors B. pneumatic hammers
   C. paint sprayers D. automobiles

3. Of the following grades of SAE crankcase oils, the one which is RECOMMENDED for year-round use is 3.____
   A. 10W-30 B. 30 C. 20W D. 10W

4. Of the following, wheel misalignment in an automobile USUALLY results in 4.____
   A. frequent stalling B. improper clutch action
   C. rapid tire wear D. impaired shock absorber action

5. Of the following, the EASIEST method of locating a defective spark plug in a gasoline engine is to 5.____
   A. take out all the spark plugs and examine them
   B. short circuit the spark plugs one at a time
   C. replace all of the spark plugs with new ones
   D. rotate all the spark plugs

6. The one of the following conditions which may cause the fuel mixture in a gasoline engine to be too rich is 6.____
   A. water in the gasoline B. a dirty air cleaner
   C. a punctured muffler D. vapor lock in the fuel line

7. If the battery of a car is constantly running dry, the one of the following items which should be checked FIRST is the 7.____
   A. generator B. ignition switch
   C. relay D. voltage regulator

8. In a gasoline engine, the throttle vale is a part of the 8.____
   A. fuel tank B. carbureto
   C. crankcase D. water radiator

9. If a car does not start on damp days, the trouble is MOST likely in the _____ system. 9.____
   A. ignition B. fuel C. lubricating D. cooling

10. The one of the following terms that applies to the relationship between the front axle and the steering mechanism of an automobile is 10.____
   A. camber B. armature C. crankshaft D. camshaft

11. The function of a carburetor on a gasoline engine is to 11.____
   A. filter the gasoline
   B. mix air and gasoline in the correct proportions
   C. pump the gasoline into the cylinder
   D. filter the air coming into the engine

12. An automotive ignition coil is used in the electrical system of a gasoline engine to 12.____
   A. reduce arcing across the breaker points
   B. transformers low voltage to high voltage
   C. operate the ignition switch
   D. charge the battery

13. The purpose of the thermostat in the cooling system of a gasoline engine is to 13.____
   A. indicate the temperature of the cooling water
   B. control water flow so as to prevent excessive pressure in the radiator
   C. prevent overheating of the cooling water
   D. prevent circulation of the cooling water when the engine is cold

14. Of the following sets of items, the BEST one to use to clean and adjust ignition points is 14.____
   A. crescent wrench, V-block, and sandpaper
   B. screwdriver, feeler gauge, and point file
   C. scraper, micrometer, and sandpaper
   D. pincers, micrometer, and emery cloth

15. The MAIN reason for not allowing oily rags to accumulate in storage closets is that 15.____
   A. a rancid odor will develop near the closet
   B. the closet will look messy
   C. oil will drip onto the floor
   D. a fire may start by spontaneous combustion

16. A certain paint can cover 310 square feet per gallon. The number of gallons of this paint required to paint 200 lines each 6 inches wide and 18 feet, 6 inches long is MOST nearly 16.____
   A. 2 B. 4 C. 6 D. 8

17. Paint brushes that are used with an oil-based paint are usually cleaned with 17.____
   A. turpentine B. linseed oil C. acetone D. alcohol

18. Assume that, while you are using an electric drill with a long electric cord, the drill suddenly stops operating. Of the following, the FIRST thing that you should do is to 18.____
    A. remove the casing of the drill to see whether the insulation of the armature is damaged
    B. check whether the cord is still plugged into the outlet
    C. check the fuses in the supply circuit
    D. inspecft the cord for a broken wire

19. A cold chisel with a "mushroomed" head is properly "dressed" by 19.____
    A. filing the cutting edge
    B. heating the head until it is red hot and quenching it in oil
    C. grinding off the turned over material
    D. heating the head of the chisel until it is red hot and, after letting it cool slowly, tapping it until all the chips fall off

20. Of the following sets of items, the BEST one to use to clean and adjust ignition points is 20.____
    A. crescent wrench, V-block, and sandpaper
    B. screwdriver, feeler gauge, and point file
    C. scraper, micrometer, and sandpaper
    D. pincers, micrometer, and emery cloth

KEY (CORRECT ANSWERS)

1.	A	11.	B
2.	D	12.	B
3.	A	13.	D
4.	C	14.	B
5.	B	15.	D
6.	B	16.	C
7.	D	17.	A
8.	B	18.	B
9.	A	19.	C
10.	A	20.	B

TEST 3

DIRECTIONS: Each question or incomplete statement is followed by several suggested answers or completions. Select the one that BEST answers the question or completes the statement. *PRINT THE LETTER OF THE CORRECT ANSWER IN THE SPACE AT THE RIGHT.*

1. Assume that, while you are using an electric drill with a long electric cord, the drill suddenly stops operating. Of the following, the FIRST thing that you should do is to 1.____
   A. remove the casing of the drill to see whether the insulation of the armature is damaged
   B. check whether the cord is still plugged into the outlet
   C. check the fuses in the supply circuit
   D. inspect the cord for a broken wire

2. A cold chisel with a "mushroomed" head is PROPERLY "dressed" by 2.____
   A. filing the cutting edge
   B. heating the head until it is red hot and quenching it in oil
   C. grinding off the turned over material
   D. heating the head of the chisel until it is red hot and, after letting it cool slowly, tapping it until all the chips fall off

3. A pipe reamer is used to 3.____
   A. thread pipe
   B. enlarge the size of a pipe
   C. remove burrs from the inside of a pipe
   D. join pipes of different sizes

4. Where only a short swing of the handle is possible, the BEST tool to use to tighten a nut or bolt is the _____ wrench. 4.____
   A. Stillson B. open end C. monkey D. ratchet

5. The wrench which is used on set screws is COMMONLY called the _____ wrench. 5.____
   A. torque B. Allen C. Stillson D. Crescent

6. A box wrench is BEST used on 6.____
   A. Allen screws B. pipe fittings
   C. hexagonal nuts D. knurled thumb screws

7. The BEST screwdriver to use when driving screws in close quarters is the 7.____
   A. butt B. angled C. Phillips D. offset

8. A "12-24" screw is MOST likely a _____ screw. 8.____
   A. machine b. sheet metal C. lag D. wood

9. The one of the following fasteners which is threaded at both ends is the 9.____
   A. lag screw B. stud
   C. bolt D. machine screw

10. Tips of masonry drills are USUALLY made of 10.____
   A. carbide B. corundum C. mild steel D. beryllium

11. A 5-inch length of pipe with male threads at each end is called a 11.____
   A. stud B. coupling C. sleeve D. nipple

12. Grade No. 2 sandpaper is 12.____
   A. finer than grade 1/0 B. coarser than grade 3
   C. finer than grade 2/0 D. coarser than grade 1

13. The one of the following lists of materials which includes ALL of the ingredients of concrete is cement, 13.____
   A. gravel, and water B. lime, sand, and water
   C. sand, gravel, and water D. sand, and water

14. The MAIN purpose of the tool known as a file card is to _____ files. 14.____
   A. clean B. sort out
   C. prevent damage to D. sharpen

15. The pull exerted by a man lifting a 200 lb. load by means of a four-part block and fall, ignoring friction, is _____ lbs. 15.____
   A. 100 B. 75 C. 50 D. 25

16. Of the following, turpentine is a solvent for 16.____
   A. shellac B. latex paint
   C. calcimine D. red lead paint

17. In a truck's gasoline engine, the condenser is a part of the 17.____
   A. distributor B. cooling system
   C. power take off D. fuel system

18. Pneumatic tools are operated by a(n) 18.____
   A. air compressor B. Pelton wheel
   C. Archimedean screw D. hydraulic ram

19. The gauge on the tank of an air compressor measures 19.____
   A. temperature of air in the tank B. pressure of air in the tank
   C. humidity of the atmosphere D. barometric pressure

20. A paint sprayer may have gauges showing the pressure of the tank, the paint pressure, and the atomizer pressure. When the sprayer is operating properly, the 20.____
   A. paint pressure is higher than the tank pressure
   B. atomizer pressure is higher than the tank pressure
   C. paint and atomizer pressures are equal
   D. atomizer pressure is higher than the paint pressure

KEY (CORRECT ANSWERS)

1.	B	11.	D
2.	C	12.	D
3.	C	13.	C
4.	D	14.	A
5.	B	15.	C
6.	C	16.	D
7.	D	17.	A
8.	A	18.	A
9.	B	19.	B
10.	A	20.	D

EXAMINATION SECTION
TEST 1

DIRECTIONS: Each question or incomplete statement is followed by several suggested answers or completions. Select the one that BEST answers the question or completes the statement. *PRINT THE LETTER OF THE CORRECT ANSWER IN THE SPACE AT THE RIGHT.*

1. A traffic sign states that parking is permitted on Sundays and Holidays. According to the traffic regulations of the city, the holiday on which parking is NOT permitted in the area covered by the sign is 1.____

   A. New Year's Day B. Memorial Day
   C. Thanksgiving Day D. Lincoln's Birthday

2. An intrastate bus is a bus that runs 2.____

   A. only in one state
   B. in 2 states only
   C. between the United States and Canada
   D. between any states in the Union

3. According to the traffic regulations of the Department of Traffic, a pedestrian facing a red signal at an intersection 3.____

   A. has the right of way over automobiles having a green signal
   B. has the right of way over trucks having a green signal
   C. may not enter the intersection facing the red signal
   D. may enter the intersection, facing the red signal, if he can do so safely without interfering with traffic

4. This sentence was taken from the traffic regulations of the City Department of Traffic with respect to yield signs: 4.____
   Proceeding past such sign with resultant collision or other impedance or interference with traffic on the intersecting street shall be deemed prima facie in violation of this regulation. The words prima facie mean MOST NEARLY

   A. probably B. possibly or likely
   C. literally or completely D. guilty

5. Where signs on city streets do not indicate otherwise, the MAXIMUM speed limit in the city is, in miles per hour, 5.____

   A. 15 B. 20 C. 25 D. 30

6. Making a U-turn in the city is NOT permissible on any 6.____

   A. street
   B. street in a residential district
   C. street in a business district
   D. 2-way street

7. A person stops his car in front of a hydrant and remains in the car. According to the traffic regulations of the City Department of Traffic, 7.____

A. this is illegal if he is within 15 feet of the hydrant
B. it is legal
C. he does not have to move if so ordered by a policeman
D. he may remain there provided he is far enough away from the hydrant so as not to interfere with hose lines

8. Taxicabs are 8.____

A. not permitted to cruise
B. permitted to cruise in residential areas only
C. permitted to cruise in business areas only
D. permitted to cruise in all boroughs except Manhattan

9. Of the following, the one that is the MAIN cause of fatal accidents is 9.____

A. direction signals not working
B. windshield wipers not working
C. improper alignment of the wheels
D. defective brakes

10. The capacity of an approach to an intersection is prinarily dependent upon 10.____

A. slope of through band
B. cycle length
C. offsets
D. through band width

11. To handle heavy traffic movements which tend to cause congestion at an intersection, it is often necessary to 11.____

A. use a standard 3-color (RAG) traffic control signal on all four corners
B. add arrow indications to traffic signals permitting movements in a certain direction when other traffic is halted
C. use 2-color instead of 3-color traffic signals
D. install a flasher caution signal facing the direction of heavy traffic flow

12. Elm Street and Oak Street are one-way streets that intersect. 12.____

A. Cars may turn either right from both streets or left from both streets depending on the direction of travel.
B. If cars may turn right into one street, they may not turn right into the other.
C. Only right turns are permitted in both streets.
D. Only left turns are permitted in both streets.

13. Of the following intersections where one street dead ends into another, the one that is SAFEST is 13.____

A. 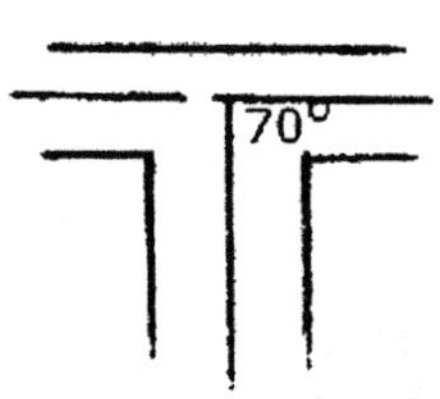

B. 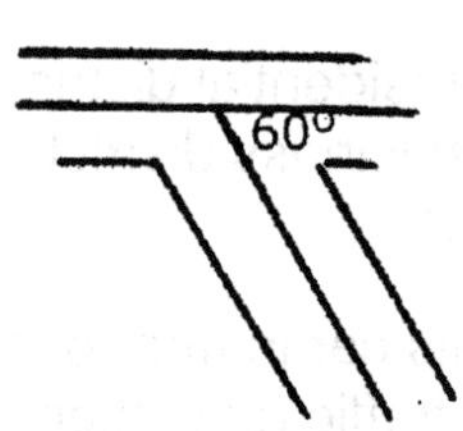

C. 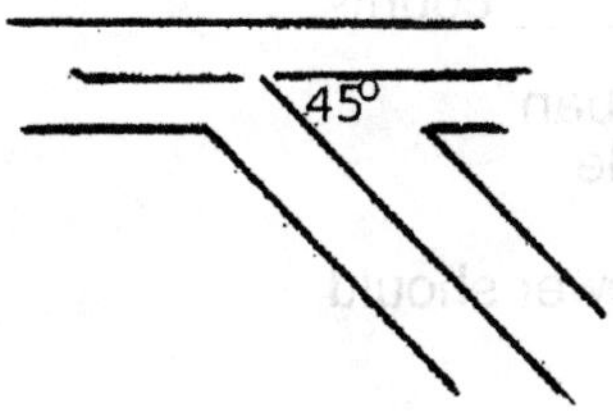

D. 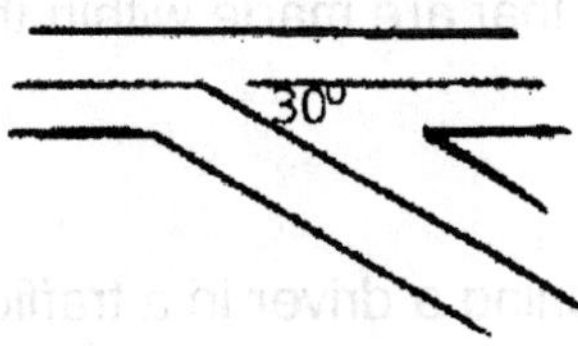

14. Driver interview, tag on vehicle, and postal cards are all methods of obtaining information relative to 14.____

    A. vehicle miles traveled
    B. accident data
    C. motor vehicle registration
    D. origin and destination

15. A study of motor vehicle volume normally includes all but one of the following: 15.____

    A. Directional movements
    B. Motor vehicle occupancy
    C. Motor vehicle classification
    D. Number of vehicles per unit of time

16. Counts made with automatic recorders must always be supplemented with manual observations to ascertain 16.____

    A. hourly distribution
    B. directional distribution
    C. vehicle classification
    D. turning movements

17. A cordon count is USUALLY made on a 17.____

    A. weekday B. Saturday C. Sunday D. holiday

18. Of the following vehicles, the one that need NOT be stopped at an origin and destination station is a 18.____

    A. bus B. foreign car
    C. station wagon D. coal truck

19. A turning movement count is USUALLY taken at 19.____

    A. a toll station B. a highway intersection
    C. a bus terminal D. the end of a highway

20. A manual traffic count is 20.____

    A. a mechanical counter tabulating pedestrians
    B. the number of manuals issued in a traffic survey
    C. an estimated volume of traffic
    D. the number of motor vehicles counted by the person assigned

21. Traffic counts that are made within the city limits are ______ counts. 21.____

A. rural
B. suburban
C. urban
D. sample

22. When questioning a driver in a traffic survey, the interviewer should 22.____

A. explain briefly the reason for the interview
B. insist on having his questions answered
C. get the signature of the person interviewed
D. report the person interviewed, if he did not cooperate

23. In gathering data for a traffic survey, it was decided to use only the period from 6:00 A.M. to 10:00 P.M. 23.____
The reason for choosing this period is MOST likely that

A. employee morale would drop if the inspectors were required to work during the night
B. the public would not cooperate during the late night or early morning hours
C. it is inconsiderate to disturb the public in the middle of the night
D. the information obtained at that time would be considered adequate

24. Of the following data, the one that is MOST significant in a traffic survey is the 24.____

A. locations between which the car travels
B. number of cars in the driver's family
C. number of drivers operating the car
D. average annual mileage of the car

25. The MAIN purpose for making a motor vehicle volume survey of a particular route is to provide basic data for determining 25.____

A. the extent of group riding
B. whether prevailing speeds are too fast for conditions
C. a plan of traffic control
D. where and how much parking space may be needed

26. Of the following studies, the one which is LEAST related and would probably NOT be included in making a traffic safety survey is 26.____

A. street and off-street parking
B. driver observance of stop signs
C. pedestrian observance of traffic signals
D. accident records and facts

27. Of the following, the one which would NOT usually require a traffic survey is 27.____

A. revision of parking time limits to assure most efficient usage of curb space
B. creation of off-street parking facilities
C. important trends in traffic characteristics and transportation demands
D. complaints from residents in a particular area on the disturbance caused by heavy traffic moving through that area

28. A *spot-map* is a graphic method which is used to 28.____

A. show types of traffic signals located at the main intersections in a community
B. analyze the distribution of accidents within a community area
C. arrive at reasonable accident rates
D. show grades, width, roadway surface, and merging traffic streams in a community

29. A survey was made for the purposes of installing traffic control signals at a certain intersection of a main street and cross street in a certain area. The survey shows that although traffic is relatively heavy during the day, it becomes very light at night. 29.____
In such a situation, it would be MOST desirable to

A. continue the full sequence of indications as in the daytime
B. continue operation of the signals, but lengthen the cycle of intervals
C. completely extinguish the signals leaving the intersection uncontrolled
D. extinguish the signals but provide a flasher mechanism on the controller

30. If the capacity of an approach to an intersection is 3600 vehicles per hour of green and the go phase on this approach is 40 seconds out of a 60-second cycle, the equivalent volume is ______ vehicles per hour. 30.____

A. 2400 B. 3600 C. 5400 D. 2000

31. If a section of a highway 10 miles long carries an annual daily traffic of 5,000 vehicles and there are two deaths in a year, the death rate is 31.____

A. 2.0 deaths per 5,000 vehicles
B. 11.0 deaths per 100 million vehicle miles
C. 11.0 deaths per million vehicle miles
D. 2.0 deaths per 50,000 vehicle miles

32. If the difference in elevation between two intersections 300 feet apart is 6 feet, the grade along the street is 32.____

A. 2% B. 2 C. 0.002 D. 6%

33. If on a highway a car passes a given point every 5 seconds, the number of cars per hour passing the given point on the highway is 33.____

A. 360 B. 480 C. 600 D. 720

34. The cost of concrete paving for a strip of driveway 50 feet long, 10 feet wide, and 6 inches deep, if concrete in place costs $30 per cubic yard, is, in dollars, MOST NEARLY (27 cubic feet = 1 cubic yard) 34.____

A. 278 B. 318 C. 329 D. 380

Questions 35-38.

DIRECTIONS: Questions 35 through 38 relate to the sketch below.

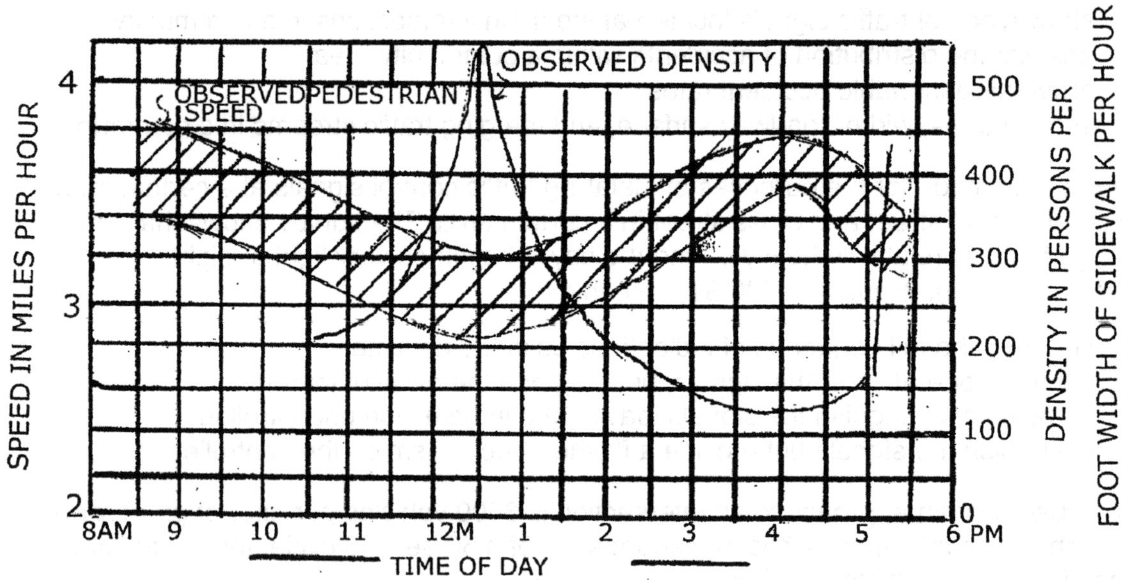

35. Assuming a 10' wide sidewalk, the number of people that would pass the given point at 12:00 M in 10 minutes is MOST NEARLY 35._____

A. 580 B. 680 C. 780 D. 880

36. At 10:00 A.M., you could expect a person to be walking at a speed 36._____

A. of 3 miles per hour
B. between 300 and 420 feet per hour
C. between 3.2 and 3.65 miles per hour
D. of 4.5 feet per second

37. The highest average number of people using the sidewalk will USUALLY occur at 37._____

A. 9 A.M. B. 12:30 P.M. C. 4 P.M. D. 5 P.M.

38. Of the following statements relating to the diagram, the one that is MOST NEARLY CORRECT is 38._____

A. the minimum walking speed observed is 2 miles per hour
B. data for the survey was taken continuously for 24 hours
C. as the number of people using the sidewalk increases, the speed at which they walk decreases
D. the minimum observed density is 300 people per hour per foot width of sidewalk

39. A vehicle moving at 30 miles per hour is moving at a speed, in feet per second, MOST NEARLY 39._____

A. 30 B. 44 C. 52 D. 60

40. A street map is to a scale 1 inch equals 600 feet. A distance of 1/2 inch on the drawing represents a distance on the ground, in feet, MOST NEARLY 40._____

A. 300 B. 600 C. 900 D. 1,200

Questions 41-42.

DIRECTIONS: Questions 41 and 42 refer to the sketches below.

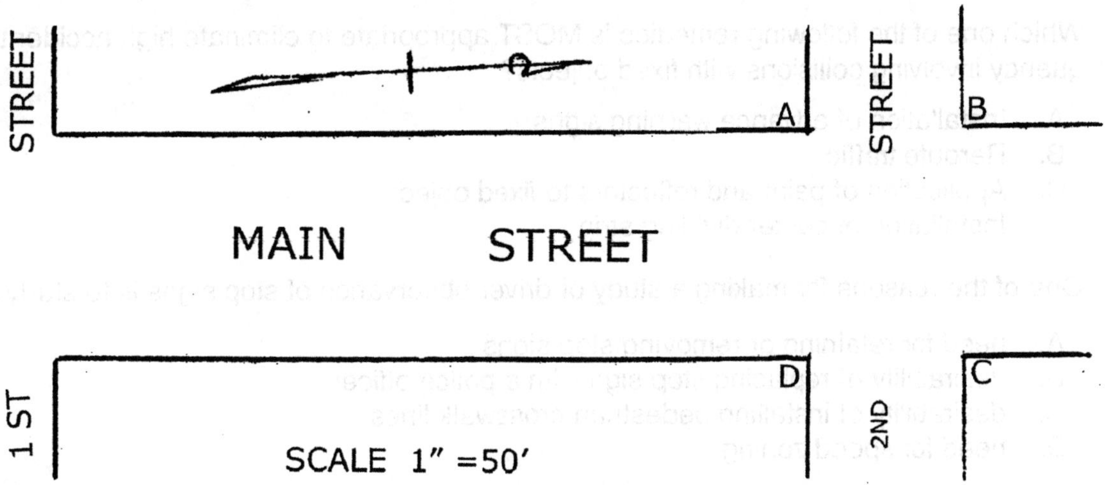

41. The length of block from 1st Street to 2nd Street is MOST NEARLY 41.____

A. 150' B. 250' C. 350' D. 450'

42. The northeast corner of Main and 2nd is 42.____

A. A B. B C. C D. D

43. The sketch shown at the right shows a right triangular island at the intersection of three streets on which is installed traffic signals A and B. Traffic conditions have increased and require than an additional traffic light be installed at point C. Electric power for signal C is to be taken from the junction box located at the base of post A and extended to C as shown by the broken line. With the distances given as shown, the length of conduit, in feet, required to extend power from A to C is MOST NEARLY 43.____

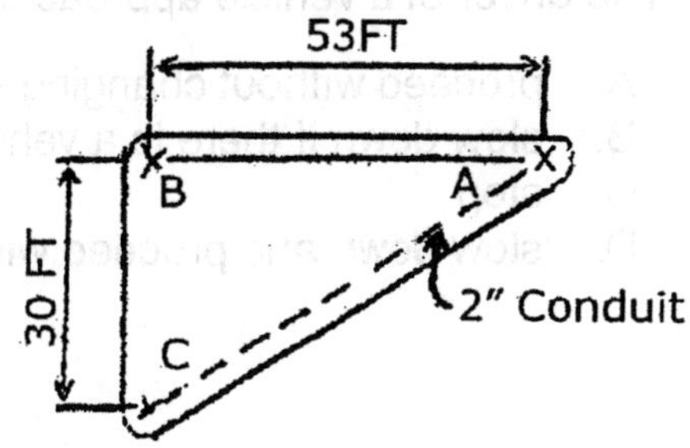

A. 44 B. 60 C. 83 D. 75

44. The volume of traffic at a certain location increased frori 1,000 to 1,500 vehicles per hour. The percentage increase of traffic is MOST NEARLY 44.____

A. 33% B. 50% C. 60% D. 40%

45. A collision diagram would MOST likely NOT show 45.____

A. direction of movement of each vehicle or pedestrian involved
B. distance of the accident to the nearest building line
C. date and hour of the accident
D. weather and road conditions

46. A graphical representation of the detailed nature of accidents occurring at a location is known as a 46.____

A. collision diagram B. condition diagram
C. accident summary D. accident spot map

47. Which one of the following remedies is MOST appropriate to eliminate high accident frequency involving collisions with fixed objects? 47.____

A. Installation of advance warning signs
B. Reroute traffic
C. Application of paint and reflectors to fixed object
D. Installation of center dividing strip

48. One of the reasons for making a study of driver observance of stop signs is to study the 48.____

A. need for retaining or removing stop signs
B. desirability of replacing stop sign with a police officer
C. desirability of installing pedestrian crosswalk lines
D. need for speed zoning

49. Which one of the following remedies is MOST appropriate to eliminate high accident frequency involving pedestrian-vehicular collisions at intersections? 49.____

A. Installation of turning guide lines
B. Installation of painted pavement lane lines
C. Installation of pedestrian cross-walk lines
D. Removal of view obstruction

50. The driver of a vehicle approaching a yield sign is required to 50.____

A. proceed without changing speed
B. slow down if there is a vehicle in the intersection
C. stop
D. slow down and proceed with caution

KEY (CORRECT ANSWERS)

1. D	11. B	21. C	31. B	41. B
2. A	12. B	22. A	32. A	42. C
3. C	13. A	23. D	33. D	43. B
4. C	14. D	24. A	34. A	44. B
5. C	15. B	25. C	35. A	45. B
6. C	16. C	26. A	36. C	46. A
7. A	17. A	27. D	37. B	47. C
8. A	18. A	28. B	38. C	48. A
9. D	19. B	29. D	39. B	49. C
10. D	20. D	30. A	40. A	50. D

TEST 2

DIRECTIONS: Each question or incomplete statement is followed by several suggested answers or completions. Select the one that BEST answers the question or completes the statement. *PRINT THE LETTER OF THE CORRECT ANSWER IN THE SPACE AT THE RIGHT.*

1. No person shall stop, stand, or park a vehicle closer to a fire hydrant than 1.____

   A. 17' B. 10' C. 15' D. 12'

2. When stopping is prohibited by signs or regulations and no conflict exists with other traffic, the driver of a vehicle is 2.____

   A. permitted to stop temporarily
   B. not permitted to stop
   C. permitted to stand
   D. permitted to park

3. Where there is a *No Parking* sign, a person may 3.____

   A. not stop his vehicle
   B. stop his vehicle to discharge passengers
   C. stop his vehicle and leave it unattended for a maximum of 10 minutes
   D. stop his vehicle and leave it unattended for a maximum of 5 minutes

4. Of the following, the MOST restrictive parking sign is 4.____

   A. no standing
   B. no parking
   C. taxi stand
   D. bus stop

5. A highway sign that is classified as a <u>Guide</u> sign is 5.____

   A. Stop
   B. No Passing
   C. Narrow Road
   D. North Bound

6. A highway sign that is classified as a <u>Warning</u> sign is 6.____

   A. No U Turn
   B. Hill
   C. Speed Limit 50
   D. Do Not Enter

7. A highway sign that is classified as a <u>Regulatory</u> sign is 7.____

   A. One Way
   B. Men Working
   C. RR
   D. Detour

8. A traffic device that has the same effect as a stop sign is a 8.____

   A. flashing yellow
   B. flashing red
   C. yield sign
   D. detour sign

9. A warrant for a certain type of traffic control device is a(n) 9.____

   A. official order to install the device
   B. application from a local community for the device
   C. reason for installing the device
   D. request to remove the device

10. Shapes of signs on state highways convey definite information. The sign to the right means 10.____

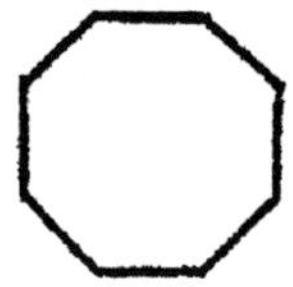

A. steep hill - slow down
B. come to a full stop
C. you may proceed with caution
D. approaching narrow bridge

11. Where flasher mechanisms must be installed at intersections of a main street and a cross street as a warning signal, it would be BEST to have flashing 11.____

A. amber on the main street and flashing red on the cross street
B. red on the main street and flashing amber on the cross street
C. red on the main street only
D. amber on the cross street only

12. The primary purpose of *progressive timing* of traffic control signals is to 12.____

A. allow the largest volume of traffic flow at the safest speed along a particular route
B. permit slow drivers to travel at an increased speed
C. permit the largest volume of pedestrian traffic to cross safely at the same time
D. reduce traveling speed so that motorists have vehicles under constant control

13. A hazard marker, for example, at the end of a dead-end street, would MOST likely be 13.____

A. yellow background with black letters
B. yellow background with red letters
C. a reflector type marker
D. a warning sign

14. Of the following, the BEST reason for having markings that are uniform in design, position, and application is that 14.____

A. less skill is required to provide the markings
B. they cost less when they are uniform
C. there is no harm done in providing them even where there is no need
D. they may be recognized and understood instantly

15. If numerous pedestrian accidents occur at a signalized intersection, a pertinent study to help evaluate the problem would be 15.____

A. signal timing
B. motor vehicle volume
C. pedestrian observance of traffic signals
D. driver observance of pedestrian right of way

16. Which one of the following types of fixed-time signal systems is MOST desirable?______ system. 16.____

A. Flexible progressive
B. Alternate
C. Simple progressive
D. Simultaneous

17. Of the following statements relating to traffic actuated signals, the one that is CORRECT is 17.____

A. it is especially useful at little used intersections
B. the length of time the green light is on is not constant
C. it can only be used at the intersection of one-way streets
D. it can only be used at the intersection of two-way streets

18. An advantage of the three lens signal (red, yellow, and green) over the two lens signal (red and green) is that it 18.____

A. enables cars within the intersection to clear
B. allows pedestrians to cross the intersection safely
C. may be operated as a traffic actuated signal
D. may be used as a caution signal when not used as a stop and go signal

19. A fixed time signal is one by which traffic stops and goes 19.____

A. for equal time periods
B. according to a predetermined time schedule
C. by manual control
D. according to the volume of traffic

20. The proper installation of vehicle detectors is MOST important for a 20.____

A. pedestrian push-button installation
B. fixed time signal system
C. traffic actuated signal
D. progressive system

21. Of the following, the one that is NOT considered a disadvantage in the use of pavement markings is they 21.____

A. may be obliterated by snow
B. may not be clearly visible when wet
C. must be used with other devices such as traffic signs or signals
D. are subject to traffic wear

22. *It is often desirable to mark lines on the pavement to indicate the limits and the clearance of the overhang on turning streetcars.* 22.____
This safety measure is NOT required in this city because

A. there are no streetcars in this city
B. city traffic is controlled by other suitable devices
C. city traffic is not fast enough to require it
D. streetcars in this city turn only at the depot and not in the streets

23. A yellow curb marking may be used at all but one of the following: 23.____

A. A fire hydrant
B. A bus stop
C. A depressed curb leading to a loading platform
D. Where parking is prohibited from 8 A.M. to 6 P.M.

24. Stop lines or limit lines are used to indicate 24.____

A. parking space limits to prevent encroachment on a fire hydrant zone
B. the marking of stalls where parking meters are used
C. the point behind which vehicles must stop in compliance with a traffic signal
D. where pedestrians are permitted to cross a street

25. An island, as applied to traffic control, 25.____

A. provides a safe area for a traffic patrolman
B. segregates pedestrians and vehicles
C. provides a clear area for a bus stop
D. establishes a barrier between opposite lanes of traffic

26. Of the following, the one which is NOT a method for providing channelization of traffic is by 26.____

A. permanent islands or strips
B. pavement markings
C. use of stanchions
D. mounting traffic signal at center of intersection

27. The PRIMARY purpose for marking the pavement of heavily traveled thoroughfares into lanes is to 27.____

A. slow up traffic
B. prevent accidents
C. speed up traffic
D. keep slow drivers on the right side of the road

28. When parking is not otherwise restricted in the city, no person shall park a commercial vehicle in excess of hours. 28.____

A. 2 B. 4 C. 3 D. 6

29. A condition which need NOT be considered in making a general parking survey is 29.____

A. reasons for parking at various locations
B. street and roadway widths and surfaces
C. average time vehicles remained at various locations
D. sidewalk obstructions, such as lamp posts and fire posts

30. Concerning the purpose of parking meters, the statement which is NOT true is 30.____

A. assist in reducing overtime parking at the curb
B. increase parking turnover
C. eliminate the need for off-street parking facilities
D. facilitate enforcement of parking regulations

31. The MOST efficient layout of parking spaces in a large lot is to place the stalls _______ to the aisles. 31.____

A. parallel
B. at right angles
C. at a 30° angle
D. at a 60° angle

32. The time limits set by cities for parking on city streets during the daytime 32.____

A. is considered strictly a policing problem
B. is shorter in concentrated business areas
C. will vary directly with the amount of traffic on the street
D. is uniform for all sections of the city

33. Four parts of a survey report are listed below, not necessarily in their proper order: 33.____

I. Body of report
II. Synopsis of report
III. Letter of transmittal
IV. Conclusions

Which one of the following represents the BEST sequence for inclusion of these parts in a report?

A. III, IV, I, II
B. II, I, III, IV
C. III, II, I, IV
D. I, III, IV, II

34. A traffic control inspector recommends that an illuminated advertising sign near a signal light be removed. 34.____
The reason for this recommendation is MOST likely that

A. a driver's attention may be attracted to the sign rather than the road
B. the similarity of colors may cause confusion
C. such signs mar the beauty of the roadside
D. the sign encroaches upon public property

35. Of the following, the MOST important value of a good report is that it 35.____

A. reflects credit upon the person who submitted the report
B. provides good reference material
C. expedites official business
D. expresses the need for official action

36. The MOST important requirement in report writing is 36.____

A. promptness in turning in reports
B. length
C. grammatical construction
D. accuracy

37. You have discovered an error in your report submitted to the main office. 37.____
You should

A. wait until the error is discovered in the main office and then correct it
B. go directly to the supervisor in the main office after working hours and ask him unofficially to correct the answer
C. notify the main office immediately so that the error can be corrected if necessary
D. do nothing, since it is possible that one error will have little effect on the total report

38. The use of *radar* by police as a means of apprehending motorists who exceed the speed limit has recently been challenged in court on the grounds that 38.____

A. the motorists are not forewarned
B. the speed limits have not been posted

C. the equipment does not give reliable results
D. there is no sworn evidence that a speed violation took place

39. Of the following, the one which is generally classified as a commercial vehicle is a 39.____

A. station wagon
B. chauffeur-driven passenger car
C. taxicab
D. truck

40. A divided arterial highway for through traffic with full or partial control of access is generally referred to as an 40.____

A. expressway
B. parkway
C. freeway
D. major street

41. Of the following, the MOST important advantage to be gained by converting a two-way north-south street to a one-way street is 41.____

A. *decrease* the number of accidents
B. *decrease* the need for bus service
C. *increase* the average speed of traffic
D. *increase* the turnover at curbs

42. Of the following, the BEST road for heavy traffic is 42.____

A. two lane
B. three lane
C. four lane undivided
D. four lane divided

43. When weekend traffic differs greatly from weekday traffic, 43.____

A. the average daily traffic figure is used in estimating weekend traffic
B. weekend traffic counts should be made as well as weekday counts
C. the traffic count for another road in the area should be used
D. traffic counts should be made at different seasons of the year

44. Work is now going on to approximately double the car-carrying capacity of which one of the following? 44.____

A. Car parkways
B. Bridges
C. Tunnels
D. HOV lanes

45. The MOST recent major change in the specifications of the federally aided highway program is 45.____

A. increasing the permissible grades or roads
B. requirements for drainage
C. lane width
D. vertical clearance under bridges

46. A recent newspaper article reported that small cars are considered a danger to the federally aided highway program.
Of the following, the one that may be considered as the reason for this danger is 46.____

A. they consume less gas providing less taxes for the highway program
B. the lanes of the new highways are too wide for these cars, disorganizing the traffic flow pattern
C. the two-car family is upsetting the estimates of traffic flow
D. foreign cars are hurting American business

47. Span-wire mountings of fixed traffic control signals is generally 47.____

A. used in the city at heavily traveled intersections
B. used in the city at intersections in isolated areas
C. not used in the city
D. used at locations where more than two streets intersect

48. A map depicting straight lines drawn from points of vehicle origin to points of vehicle destination is known as ______ map. 48.____

A. desire line
B. traffic flow
C. bar
D. pie

49. Brake reaction time for most people is APPROXIMATELY ______ seconds. 49.____

A. 0.6 B. 2.0 C. 0.1 D. 1.4

50. Trucks should travel along prescribed truck routes if their overall length is equal to or exceeds 50.____

A. 27' B. 41' C. 30' D. 33'

KEY (CORRECT ANSWERS)

1. C	11. A	21. C	31. B	41. C
2. B	12. A	22. A	32. B	42. D
3. B	13. C	23. D	33. C	43. B
4. A	14. D	24. C	34. B	44. D
5. D	15. C	25. B	35. C	45. D
6. B	16. A	26. D	36. D	46. A
7. A	17. B	27. C	37. C	47. C
8. B	18. D	28. C	38. C	48. A
9. C	19. B	29. D	39. D	49. A
10. B	20. C	30. C	40. A	50. D

EXAMINATION SECTION
TEST 1

DIRECTIONS: Each question or incomplete statement is followed by several suggested answers or completions. Select the one that Best answers the question or completes the statement. *PRINT THE LETTER OF THE CORRECT ANSWER IN THE SPACE AT THE RIGHT.*

Questions 1-4.

DIRECTIONS: Answer Questions 1 to 4 based on the information given in the traffic volume table below.

TRAFFIC VOLUME COUNTS

Time (A.M.)	Main Street Northbound	Main Street Southbound	Cross Street Eastbound	Cross Street Westbound
7:00- 7:15	100	100	70	60
7:15- 7:30	110	100	80	70
7:30- 7:45	150	140	110	100
7:45- 8:00	170	160	140	130
8:00- 8:15	210	190	120	110
8:15- 8:30	180	170	90	80
8:30- 8:45	160	140	70	60
8:45-9:00	150	160	70	50
9:00- 9:15	140	150	50	50
9:15- 9:30	130	120	40	20
9:30- 9:45	120	110	30	30
9:45-10:00	120	100	30	30

1. The hour during which traffic, moving in both directions on Main Street, reached its *peak* was 1.____

   A. 7:30 - 8:30
   B. 7:45 - 8:45
   C. 8:00 - 9:00
   D. 8:15 - 9:15

2. The hour during which traffic volume, moving in both directions on Cross Street, reached its *peak* was 2.____

   A. 7:30 - 8:30
   B. 7:45 - 8:45
   C. 8:00 - 9:00
   D. 8:15 - 9:15

3. The HIGHEST average hourly volume over the three-hour period 7:00 to 10:00 was recorded for 3.____

   A. Main Street northbound
   B. Main Street southbound
   C. Cross Street eastbound
   D. Cross Street westbound

4. The *peak* 15-minute traffic volume for all directions of travel occurred between 4.____

   A. 7:30 - 7:45
   B. 7:45 - 8:00
   C. 8:00 - 8:15
   D. 8:15 - 8:30

5. Which of the following statements relating to one-way streets is CORRECT? One-way streets 5.____

A. increase turning movement conflicts between vehicles
B. decrease street capacity
C. decrease accident hazards for pedestrians
D. make it impossible to time traffic signals to control speeds

Questions 6-11.

DIRECTIONS: Answer Questions 6 to 11 based on the information given in Figure 1 below.

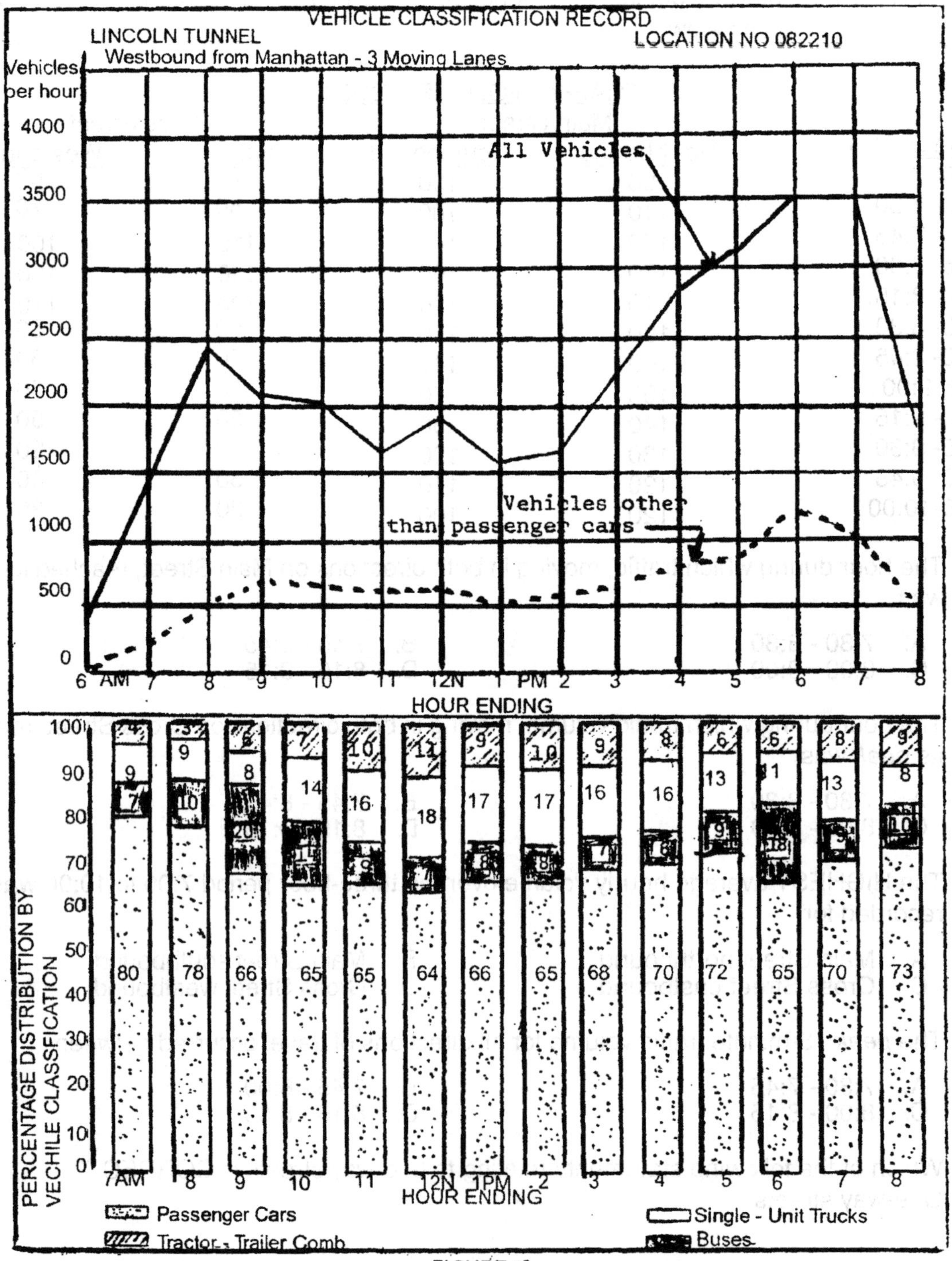

FIGURE 1

6. The total number of all vehicles traveling through the Lincoln Tunnel westbound from Manhattan between the hours of 6 A.M. and 12 Noon is *most nearly* 6.____

A. 5,500 B. 7,500 C. 9,500 D. 11,500

7. The number of passenger cars recorded during the hour ending at 7 P.M. was *most nearly* 7.____

A. 235 B. 1160 C. 2450 D. 3500

8. Excluding passenger cars, the AVERAGE number of vehicles per moving lane recorded during the peak hour was *most nearly* 8.____

A. 420 B. 1180 C. 1250 D. 3550

9. The percentage of buses recorded between 6 A.M. and 8 P.M. ranged between 9.____

A. 3% and 11%
B. 8% and 18%
C. 6% and 20%
D. 64% and 80%

10. During the study period, the percentage of single unit trucks *exceeded* the percentage of buses for _______ hours. 10.____

A. 4 B. 5 C. 9 D. 10

11. For all vehicles recorded, the recorded traffic volume during the morning peak hour was *most nearly* _______ of the volume during the evening peak hour. 11.____

A. 40% B. 50% C. 60% D. 70%

12. In urban areas, traffic volume is usually LOWEST during the month of 12.____

A. January B. March C. August D. October

13. In urban shopping areas, the *peak* traffic activity USUALLY occurs during 13.____

A. Monday afternoon and Friday night
B. Friday night and Saturday afternoon
C. Thursday night and Saturday afternoon
D. Monday night and Friday night

14. In the metric system, the unit that is closest to a mile is a 14.____

A. centimeter
B. liter
C. millimeter
D. kilometer

Questions 15-16.

DIRECTIONS: Questions 15 and 16 refer to the diagram at the top of the following Page 4.

15. Vehicle X in the diagram is heading in which direction? 15.____

A. Southeast
B. Southwest
C. Northeast
D. Northwest

16. If Vehicle X in the diagram makes a right turn at the intersection, it will be headed 16.____

A. southeast
B. southwest
C. northeast
D. northwest

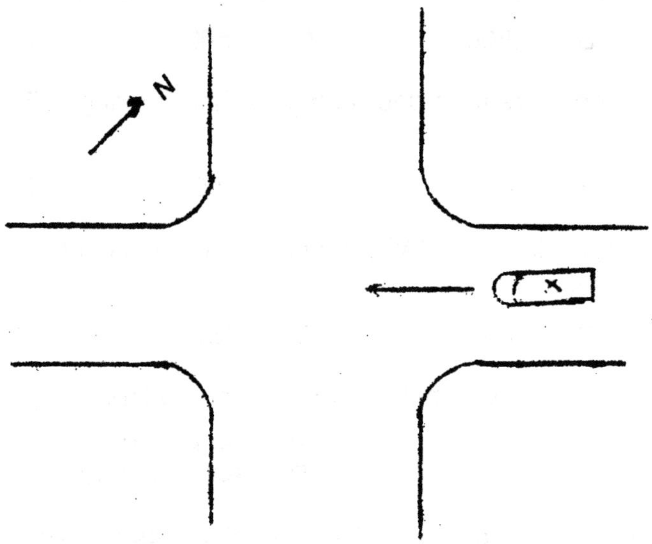

17. The one of the following that is NOT a function of channelization is 17.____

    A. control the angle of conflict
    B. favor certain turning movements
    C. protect pedestrians
    D. increase the pavement area within an intersection

18. The time of display of the yellow signal indication following the green signal indication is called the 18.____

    A. clearance interval B. time cycle
    C. traffic phase D. interval sequence

19. A lane constructed for the purpose of allowing vehicles entering a highway to increase speed to a rate that is safe for merging with through traffic is called a(n) ______ lane. 19.____

    A. auxiliary B. through
    C. acceleration D. deceleration

20. A traffic volume count which records the number and types of vehicles passing a given point is called a ______ count. 20.____

    A. rate-of-flow B. capacity
    C. classification D. roadway

21. On highways, the MAIN purpose served by barriers between traffic going in opposite directions is to 21.____

    A. stop cars if they get out of lane
    B. minimize the glare from oncoming cars
    C. prevent cars from overturning if they have blowouts
    D. prevent head-on accidents

22. Control count stations are USUALLY used to 22.____

A. establish seasonal and daily traffic volume characteristics
B. make short manual traffic counts
C. classify traffic
D. count traffic on weekends only

23. The MAIN purpose of off-center traffic lanes is to 23.____

A. protect slow-moving traffic from the hazards of fast-moving traffic
B. permit the use of special traffic control
C. provide additional capacity in one direction of travel
D. provide a slow-down area for disabled vehicles

24. Reserved transit lanes are used to 24.____

A. make sure buses stop at the curb
B. reduce bus and passenger car accidents
C. decrease transit travel times by reducing friction between buses and other vehicles
D. make it easier for people to get on and off buses

25. The slope or grade between points X and Y shown in the diagram below is 25.____

A. 4% B. 10% C. 25% D. 50%

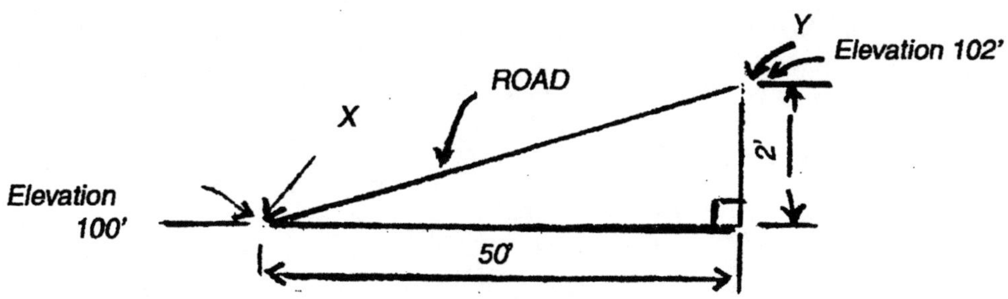

KEY (CORRECT ANSWERS)

1. B
2. A
3. A
4. C
5. C
6. D
7. C
8. A
9. C
10. C
11. D
12. A
13. B
14. D
15. B
16. D
17. D
18. A
19. C
20. C
21. D
22. A
23. C
24. C
25. A

TEST 2

DIRECTIONS: Each question or incomplete statement is followed by several suggested answers or completions. Select the one that BEST answers the question or completes the statement. *PRINT THE LETTER OF THE CORRECT ANSWER IN THE SPACE AT THE RIGHT.*

1. In the city, when parking is not otherwise restricted, commercial vehicles can park 1.____

   A. up to a maximum of one hour
   B. up to a maximum of three hours
   C. up to a maximum of eight hours
   D. without a time limitation

2. In the city, with respect to loading an parking, commercial vehicles are allowed to 2.____

   A. load or unload merchandise expeditiously in a no-standing zone
   B. park for one hour in a no-parking zone
   C. load or unload merchandise expeditiously in a no-parking zone
   D. park for one hour in a no-standing zone

3. On the Federal national highway system, highways ending in an even number run 3.____

   A. in the east-west direction
   B. both east-west or north-south
   C. in the north-south direction
   D. around cities and not through them

4. The *current* maximum allowed speed limit on Federal interstate highways is _____miles per hour. 4.____

   A. 50 B. 55 C. 60 D. 65

5. In the city, when a vehicle is too long for a single parking meter space, the vehicle may 5.____

   A. not be parked in the parking meter area
   B. be parked using more than one space but a coin must be deposited in the meter designated for each space occupied
   C. be parked using more than one space and a coin must be deposited only in the forward parking meter
   D. be parked using more than one space and a coin must be deposited only in the rear parking meter

6. In the city, some signs indicate that stopping, standing, or parking regulations are in effect every day except Sundays. Where this sign is used, stopping, standing, or parking regulations would apply on 6.____

   A. Washington's Birthday B. Brooklyn Day
   C. Columbus Day D. Election Day

7. In the city, unless signs are posted indicating specific hours during which play street regulations are in effect, such regulations are in effect on designated streets FROM 7.____

   A. 7 A.M. until 4 P.M.
   B. 8 A.M. until 1/2 hour before sunset

C. 8 A.M. to 1/2 hour after sunset
D. 8 A.M. to 8 P.M.

8. When preparing to make a turn while driving a vehicle on a roadway, a driver should signal his intention to turn AT LEAST _______ feet in advance of the turn. 8.____

A. 50 B. 100 C. 150 D. 200

9. Unless otherwise permitted or prohibited by posted signs, the MAXIMUM continuous period during which a vehicle may be parked on any roadway in the city is hours. 9.____

A. 8 B. 12 C. 24 D. 48

10. In the city, commercial vehicles may angle stand or angle park in 10.____

A. any area where no parking signs are installed, provided the street is wide enough to allow the vehicle to park at an angle
B. on any one-way street where standing is not prohibited, provided the street is wide enough to allow the vehicle to park at an angle
C. on a two-way street in areas authorized by signs, provided that the vehicle shall not occupy more than a parking lane plus one moving lane
D. on a two-way street in areas authorized by signs, provided that the vehicles shall not extend more than 10 feet from the curb

11. Which of the following is MOST restrictive to drivers of passenger cars? 11.____

A. Regulations relating to parking in front of fire hydrants
B. No parking regulations
C. No standing regulations
D. No stopping regulations

12. The MAXIMUM permitted speed limit in the city, unless signs indicate otherwise, is _______ mph. 12.____

A. 25 B. 30 C. 35 D. 40

13. With regard to right-of-way at an intersection that is NOT controlled by a traffic control device, the one of the following statements that is CORRECT is 13.____

A. the car on your right has the right-of-way
B. the car on your left has the right-of-way
C. a car preparing to enter the intersection has the right-of-way over a car in the intersection
D. a car turning left has the right-of-way over a vehicle going straight ahead

14. At an intersection controlled by traffic signals, a red arrow pointing to the right means that a right turn may 14.____

A. be made after coming to a full stop
B. be made providing the driver yields the right-of-way to all other vehicles and pedestrians
C. not be made during the period that the red arrow is illuminated
D. be made only if there is another indication showing a round green signal light

15. A flashing red traffic signal has the SAME meaning as a 15.____

A. stop sign
B. yield sign
C. flashing yellow traffic signal
D. hazardous intersection warning sign

16. Traffic signals are MOST frequently installed to reduce ______ collision accidents. 16.____

A. right-angle
B. rear-end
C. side-swipe
D. head-on

17. The CORRECT color combination for warning signs is 17.____

A. yellow lettering or symbols on a black background
B. white lettering or symbols on a red background
C. black lettering or symbols on a yellow background
D. black lettering or symbols on a white background

18. A PROGRESSIVELY timed traffic signal system will 18.____

A. turn all the signals red or green at the same time
B. usually increase the number of rear-end accidents but reduce the number of right-angle accidents
C. make it more hazardous for pedestrians to cross at the signalized intersections
D. decrease the number of stops traffic is required to make

19. The EFFECT of traffic signals on accidents is that traffic signals 19.____

A. always decrease accidents
B. sometimes increase accidents
C. never increase accidents
D. have no real effect on accidents

20. With respect to traffic devices, which of the following situations should receive the LOWEST priority in terms of repair or replacement? 20.____

A. Inoperative or malfunctioning traffic signals at an intersection
B. Missing "No Standing - Rush Hour" regulation signs
C. Missing "Yield" signs controlling the intersection of a minor street with a major street
D. Inoperative parking meters along one block in a retail shopping area

21. Of the following, the BEST reason why a stop sign would be used instead of a yield sign to control traffic at an intersection is 21.____

A. there are a larger number of rear-end accidents on the street being controlled
B. the street being controlled is less than 36 feet wide
C. visibility is limited at the intersection
D. the approaches to the intersection are offset to each other

22. The USUAL color combination used on interstate signs is ______ lettering and symbols on a ______ background. 22.____

A. white; green
B. green; white
C. white; black
D. black; white

23. The geometrical shape of a railroad crossing sign is that of a(n) 23.___

A. octagon B. circle C. rectangle D. triangle

24. The STANDARD pedestrian walking speed used in timing pedestrian signals is ______ per second. 24.___

A. 1 foot B. 4 feet C. 8 feet D. 12 feet

25. A driver approaching an intersection where a sign authorizes a right turn on a red traffic signal indication may make such a turn AND 25.___

A. has the right-of-way over all vehicles in the intersection
B. must yield right-of-way to all vehicles and pedestrians within the intersection
C. must yield right-of-way only to vehicles and pedestrians on the cross street
D. has the right-of-way over other turning vehicles

KEY (CORRECT ANSWERS)

1. B
2. C
3. A
4. D
5. C
6. B
7. C
8. B
9. C
10. C
11. D
12. B
13. A
14. C
15. A
16. A
17. C
18. D
19. B
20. D
21. C
22. A
23. B
24. B
25. B

EXAMINATION SECTION
TEST 1

DIRECTIONS: Each question or incomplete statement is followed by several suggested answers or completions. Select the one that BEST answers the question or completes the statement. *PRINT THE LETTER OF THE CORRECT ANSWER IN THE SPACE AT THE RIGHT.*

1. Under conditions of rain and fog, headlights MUST be turned on if visibility is *less than* 1.____

   A. 100 feet B. 500 feet
   C. 1,000 feet D. 1/4 of a mile

2. If a blowout occurs while a vehicle is moving, the driver SHOULD 2.____

   A. hold tightly onto the steering wheel and immediately apply steady foot pressure to the brake pedal
   B. hold tightly onto the steering wheel and turn off the roadway as soon as the blowout occurs, stopping the vehicle on the shoulder area
   C. start tapping the brake pedal with his foot, sound his horn to warn others, and move into the slow-moving lane immediately
   D. hold tightly onto the steering wheel, steer straight ahead, and ease up on the accelerator

3. Which of the following is NOT a correct action to take after parking on a shoulder of a highway? 3.____

   A. Turn on the emergency lights
   B. Have all occupants stay inside the vehicle
   C. Open the hood of the car
   D. Fasten a white cloth to the door handle or radio antenna

4. It is officially recommended that drivers stay behind cars in front of them a distance of at least one car length for every ten miles per hour of speed. 4.____
   The PRINCIPAL reason for this is to

   A. increase roadway capacity
   B. make it easier for cars to change lanes
   C. allow for enough distance to stop safely if the car ahead stops suddenly
   D. keep all the cars moving at the same speed

5. To *minimize* the glare from lights of oncoming cars at night, a driver should 5.____

   A. shift his eyes to the lower right side of his traffic lane
   B. blink his eyes frequently
   C. shift his eyes to the upper right side of his traffic lane
   D. use his upper beams to offset the glare

6. The number of yards in a mile is 6.____

   A. 5,280 B. 1,760. C. 880 D. 440

7. Vehicle classification data is MOST important in calculating the 7.____

A. capacity of a roadway
B. timing of traffic signals
C. turnover of parking in a lot
D. average speed of travel

8. Condition diagrams show 8.____

A. the same information as collision diagrams plus information about vehicle classification
B. the same information as collision diagrams plus information about vehicle speeds and traffic volumes
C. traffic volumes only
D. existing physical features at a location

9. Thirty miles per hour is *equivalent* to ______ feet per second. 9.____

A. 30 B. 44 C. 60 D. 80

10. Origin and destination studies are used CHIEFLY to 10.____

A. obtain information on travel habits
B. collect traffic volume data
C. estimate travel time between cities
D. determine roadway capacities

11. The BEST way to determine the number of cars parked in an off-street parking lot during a 12-hour period is to conduct a(n)______ study. 11.____

A. parking occupancy
B. vehicle classification
C. origin/destination
D. parking turnover

12. The BEST way to determine the number of vehicles that have been parked for more than 1 hour within a 1-hour parking meter area is to record the ______ every hour. 12.____

A. color of each vehicle parked
B. license numbers of every parked vehicle
C. make and year of each car parked
D. number of violations shown on the meters

13. A traffic flow map is used to show 13.____

A. speeds along a highway in both directions and at intersections
B. the available capacity on a highway during peak hours
C. the traffic volumes that pass through an intersection or travel along a highway
D. the one- and two-way street patterns in an area

14. The *running speed* on a highway is the 14.____

A. posted speed limit
B. length of the highway divided by the time it takes to travel the highway
C. speed for which the highway was designed
D. speed at a specific point along the highway as determined through a radar speed study

15. The 85 percentile speed on a given stretch of highway for a certain period of time is 15.____

A. the speed below which 85% of all traffic travels
B. the speed above which 85% of all traffic travels
C. 85% of the posted speed limit
D. 85% of the running speed

16. The *modal* speed on a highway is 16.____

A. average speed traveled by vehicles using the highway
B. speed value which is halfway between the highest and lowest speed recorded in a speed study
C. average spot speed at a given station on the highway
D. speed value occurring most frequently as recorded in a speed study

17. For a given number of lanes in a roadway, the capacity of the roadway 17.____

A. decreases as lane widths decrease
B. increases as lane widths decrease
C. is not affected by lane widths
D. is affected by lane widths only on steep grades

18. Speed and delay studies are used to 18.____

A. determine the number of vehicles traveling above and below the speed limit
B. establish speed limits
C. identify locations where curb parking needs to be restricted
D. measure the effectiveness of changes in traffic signal timing

19. The average annual daily traffic on a highway is the 19.____

A. total yearly volume divided by the number of days in the year
B. average weekday volume times 365
C. average of the highest and lowest 24-hour volumes recorded during the year
D. average 24-hour volume recorded exclusive of Saturdays, Sundays, and holidays

20. Turning movement counts at intersections are USUALLY made 20.____

A. with the use of mechanical counters
B. with the use of radar detectors
C. by manual means
D. by estimation based upon traffic flow diagrams

21. Under ideal roadway and traffic conditions, the basic capacity for uninterrupted traffic flow conditions foreach lane of a multi-lane roadway is ______ passenger cars per hour. 21.____

A. 500 B. 1,000 C. 2,000 D. 4,000

22. For heavy volumes of mixed traffic, the IDEAL lane width is ______ feet. 22.____

A. 10 B. 12 C. 14 D. 16

23. When describing highway capacity under various traffic conditions, flow, volumes, and speeds, levels of service definitions are used. 23.____
The level of service which describes a condition on the roadway of free flow, low volume, and high speed is known as level of service

A. A B. B C. C D. D

24. The MOST efficient use of space in a rectangular or square-shaped parking lot can USUALLY be arrived at through the use of ______ parking stalls. 24.____

A. parallel
B. 45-degree
C. 90-degree
D. a combination of angle and 90-degree

25. If a single mechanical traffic counter is installed on each approach to an intersection, the data collected will NOT include 25.____

A. peak hour volumes
B. the total volumes through the intersection
C. turning movements
D. motorcycle traffic

KEY (CORRECT ANSWERS)

1. C
2. D
3. B
4. C
5. A

6. B
7. A
8. D
9. B
10. A

11. D
12. B
13. C
14. B
15. A

16. D
17. A
18. D
19. A
20. C

21. C
22. B
23. A
24. C
25. C

TEST 2

DIRECTIONS: Each question or incomplete statement is followed by several suggested answers or completions. Select the one that BEST answers the question or completes the statement. *PRINT THE LETTER OF THE CORRECT ANSWER IN THE SPACE AT THE RIGHT.*

Questions 1-5.

DIRECTIONS: Questions 1 through 5, inclusive, refer to Figures 1 and 2, which appear below and on the following page.

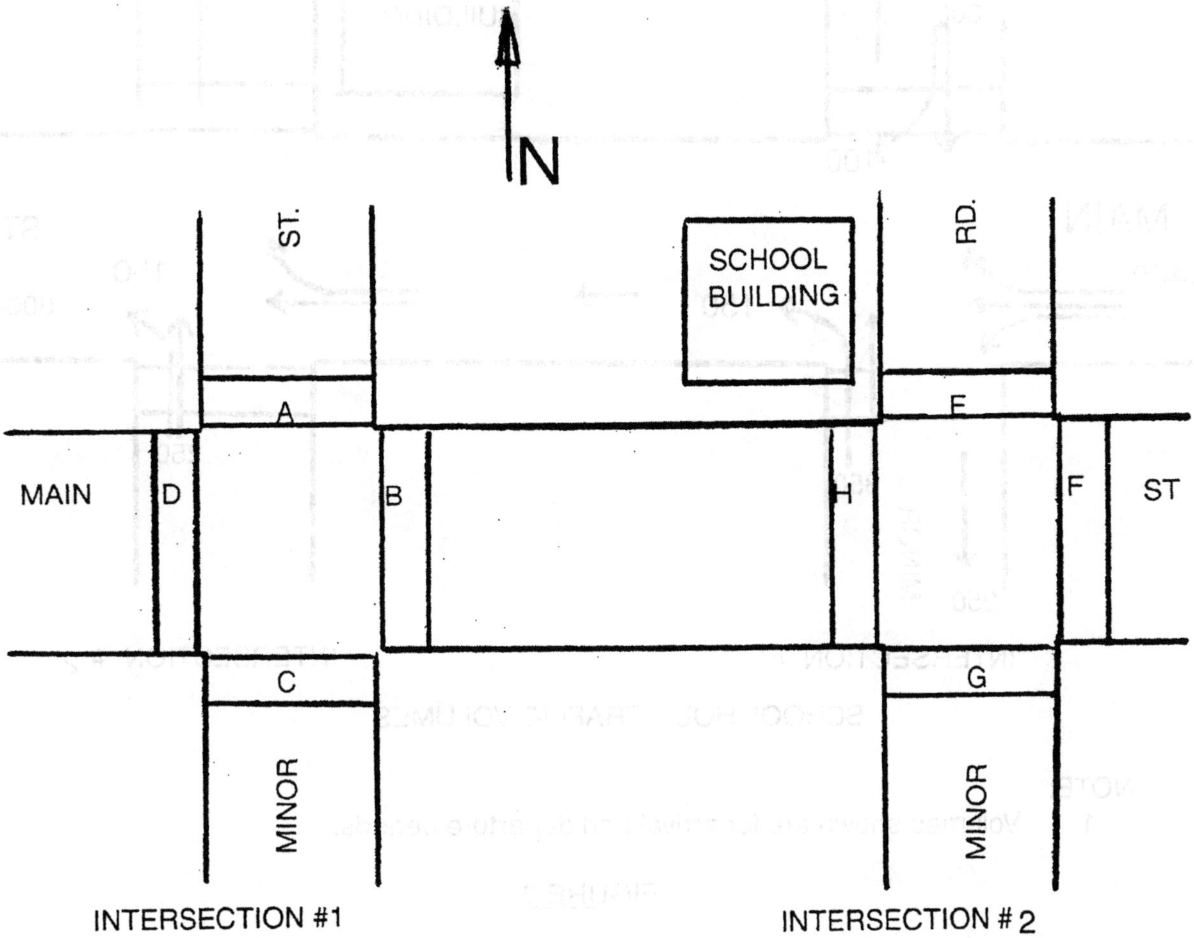

POSSIBLE CROSSWALK LOCATION

NOTES	STREET DIRECTIONS
1	Main Street Is one-way eastbound
2	Cross Street Is two-way
3	Cross Road is one-way northbound
4	Both intersections are uncontrolled

FIGURE 1

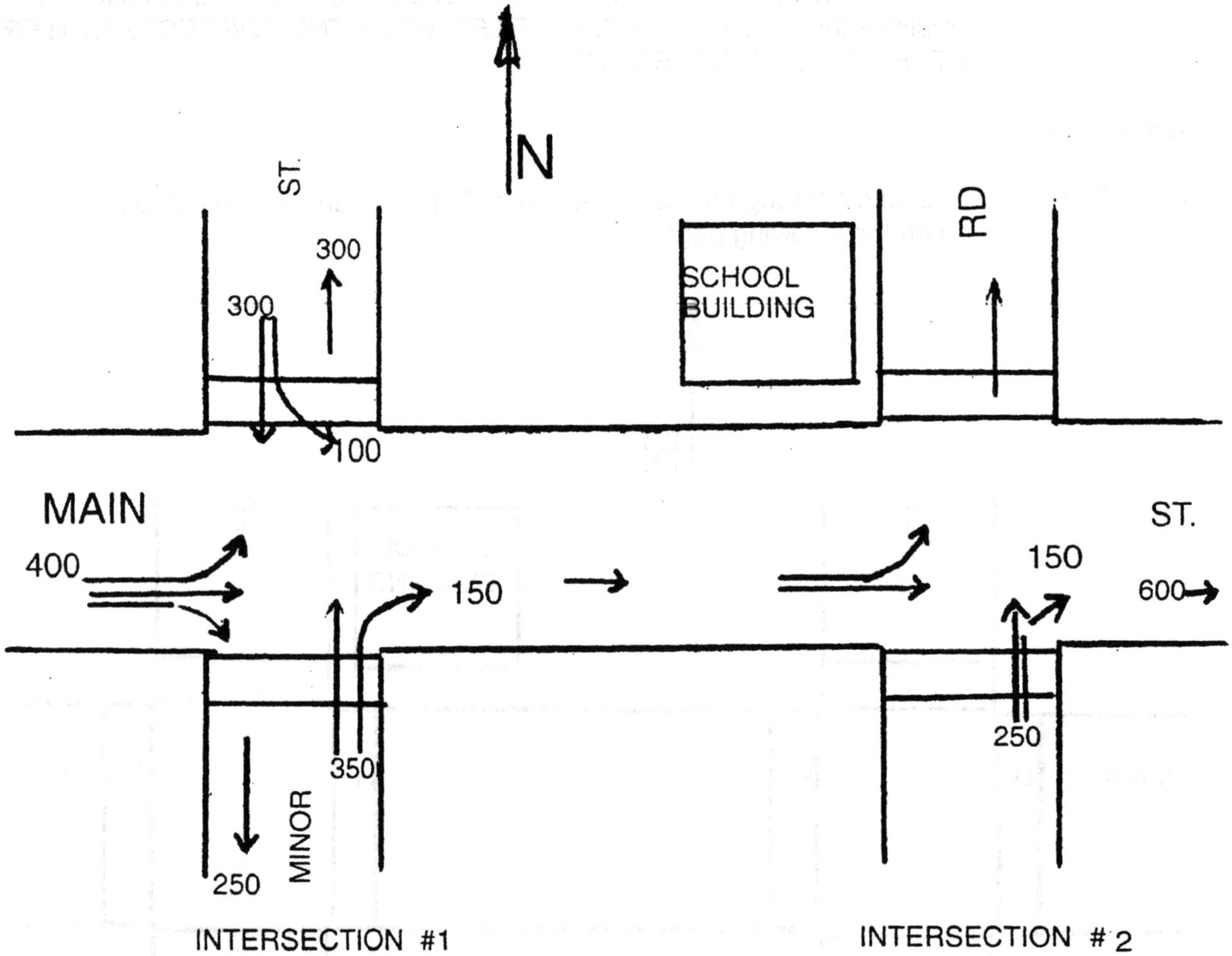

SCHOOL HOUR TRAFFIC VOLUMES

NOTE:
1 Volumes shown are for arrival and departure periods.

FIGURE 2

1. At intersection #1, the TOTAL traffic volume that crosses crosswalk B is MOST NEARLY 1.___
   A. 250 B. 350 C. 400 D. 500

2. At intersection #2, the TOTAL traffic volume that moves straight ahead at crosswalk G is MOST NEARLY 2.___
   A. 100 B. 150 C. 200 D. 250

3. At intersection #1, the TOTAL traffic volume that moves straight ahead at crosswalk D is MOST NEARLY 3.___
   A. 150 B. 200 C. 250 D. 300

4. At intersection #2, the crosswalks that should be designated as school crossings are 4.____

A. E, F, G, H
B. E. F, G
C. E, F, H
D. E, H, G

5. Assuming that only one police officer or school crossing guard can be assigned for school crossing patrol duty, the officer or guard should be assigned to intersection 5.____

A. #1 during student arrival periods and at intersection #2 during student departure periods
B. #2 during student arrival periods and at intersection #1 during student departure periods
C. #1 during both arrival and departure periods
D. #2 during both arrival and departure periods

6. In the city, the CLOSEST a car may be parked to a hydrant is ______ feet. 6.____

A. 5
B. 10
C. 15
D. 20

7. Of the following violations, the one which would NOT be recorded as a penalty on a driver's license is 7.____

A. failure to stop at a stop sign
B. double parking
C. front or rear lights not working
D. passing a red light

8. The current trend in the manufacture of new automobiles in the United States is to 8.____

A. give the new automobiles capacity for higher speeds
B. make them smaller
C. restore the running board
D. give them disappearing front lights

9. Of the following, the statement relating to parking meter spaces adjacent to fire hydrants that is MOST NEARLY correct is they 9.____

A. may be made shorter than others in the block
B. must be at least 25 feet long
C. cannot be closer than 20 feet to the hydrant
D. may be within 10 feet of the hydrant

10. In the city, parking signs that prohibit parking are made with ______ letters on a ______ background. 10.____

A. green; white
B. white; green
C. black; white
D. red; white

11. A driver whose car is parked for 8 hours in an off-street facility where the rate is 50 cents an hour for the first 5 hours and 75 cents an hour thereafter would pay 11.____

A. $6.00
B. $5.75
C. $4.75
D. $4.00

12. To encourage shoppers and other short-term parkers and to discourage commuters from using parking garages in the city, it would be BEST to 12.____

A. charge a uniform high hourly rate all day
B. charge a high rate for the first three to five hours and decrease the rate thereafter
C. charge a lower rate for the first three hours and increase the rate sharply thereafter
D. limit all parking to one-half hour

13. If an investigation of insufficient parking for customers at a busy post office revealed that the only six available spaces were occupied by all-day parkers, the recommended action should be to 13.____

A. install two-hour parking signs
B. make the area a No Parking zone
C. do nothing because the spaces are being used
D. install 20-minute meters

14. A street which can accommodate 40 parked trucks along both curbs is experiencing congestion problems because there are not enough lanes for through traffic. A survey reveals stores and businesses along both sides of the street and truck parking along both curbs. The total number of vehicles parked is never more than 20.
The recommended action should be to 14.____

A. prohibit truck parking at all times along one curb
B. prohibit truck parking for the first half of the day along one curb and the second half of the day along the other curb
C. prohibit parking on alternate days along each curb
D. establish loading zones mid-block along each curb

15. An off-street parking garage where the driver parks his own vehicle is called a _______ garage. 15.____

A. self-parking
B. ramp
C. commuter
D. mechanical

16. Off-street garages and lots where attendants park vehicles need adequate reservoir (storage) space at the entrance PRIMARILY to 16.____

A. reduce customer waiting time when picking up cars
B. reduce the number of attendants needed to park cars
C. avoid spill-back of cars into the street system
D. have extra space for parking cars when the garage fills up

17. Parking turnover is defined as the 17.____

A. capacity of a parking lot or garage divided by the number of cars parked in it
B. average number of times a parking space in a parking lot or garage is used during a given period of time
C. number of empty spaces in a parking lot or garage
D. number of space hours used during a day in a parking lot or garage

18. One-half hour parking meters would BEST serve customers of a(n) 18.____

A. supermarket
B. medical building
C. bank
D. office building

19. An off-street parking facility at a shopping center is operating at its BEST efficiency when it is _______ full 19._____

A. 100% B. 85% C. 75% D. 50%

20. 20._____

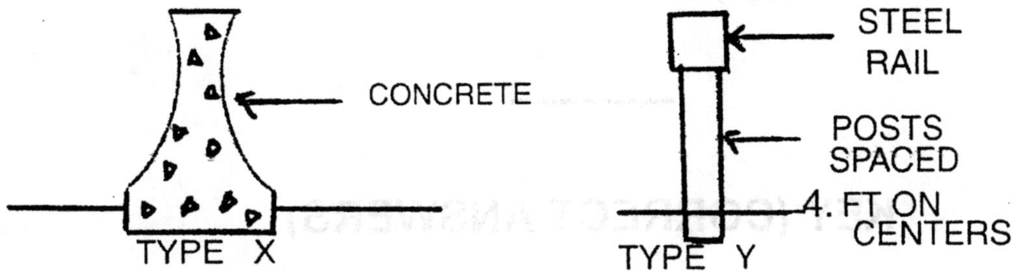

Two types of barriers are shown above, Type X and Type Y. An *advantage* of Type X barrier over Type Y barrier is that Type X barrier _______ than Type Y.

A. has a lower initial cost
B. is easier to install
C. requires less maintenance
D. is more visible

21. A fatality is MOST likely to occur in a _______ accident. 21._____

A. rear-end B. right-angle
C. side-swipe D. head-on

22. Most accidents USUALLY occur 22._____

A. during the morning rush hours
B. at midday
C. in the late afternoon and early evening
D. between midnight and dawn

23. For the United States as a whole, studies have shownthat alcohol was a contributing factor in _______ of the fatal accidents. 23._____

A. 5% B. 15% C. 25% D. 50%

24. The MAIN advantage of a red, yellow, and green light over a red and green light is that the red, yellow, and green light 24._____

A. is less expensive
B. is easier to install
C. gives the driver warning of a change in signals from green to red
D. gives the police officer firm evidence if he wants to issue a violation for passing a light

25. Dividing the total number of accidents occurring in one year on a roadway by the length of the roadway in miles will yield the 25.____

A. fatality rate for the roadway
B. accident rate per annual vehicle miles traveled
C. accident exposure rate for the roadway
D. accident rate per mile per year

KEY (CORRECT ANSWERS)

1. D
2. A
3. C
4. D
5. C

6. C
7. B
8. B
9. A
10. D

11. C
12. C
13. D
14. B
15. A

16. C
17. B
18. C
19. B
20. C

21. D
22. C
23. D
24. C
25. D

EXAMINATION SECTION
TEST 1

DIRECTIONS: Each question or incomplete statement is followed by several suggested answers or completions. Select the one that BEST answers the question or completes the statement. *PRINT THE LETTER OF THE CORRECT ANSWER IN THE SPACE AT THE RIGHT.*

1. The BEST reason, among the following, for obtaining a written and signed statement of the testimony of a witness is that 1.____

   A. unless reduced to writing, it cannot ultimately be placed in evidence in court
   B. the witness may be unavailable at the time of a subsequent trial or may attempt to change his testimony
   C. the investigator's notes of the interview may be defective or incomplete
   D. such a written statement becomes *best evidence* whereas the investigator's report is mere hearsay.

2. When an investigator hears an important statement made by a witness and the witness is not willing to reduce the statement to writing, the MOST advisable of the following procedures for the investigator to follow is to 2.____

   A. write it himself and have the witness sign it if he is willing to do so
   B. write it himself and insist that the witness sign it
   C. write it himself, making sure the witness does not see it
   D. threaten to write it himself if the witness will not do so

3. Suppose that you are interviewing an eleven-year-old boy. The CHIEF point, among the following, for you to keep in mind is that a child, as compared with an adult, is generally 3.____

   A. more likely to attempt to conceal information
   B. a person of lower intelligence
   C. more garrulous
   D. more receptive to suggestive questions

4. Of the following, witnesses to the same event should be interviewed 4.____

   A. *together* so that each can help to refresh the recollection of the others
   B. *together* so that discrepancies in their statements can be corrected more readily
   C. *separately* since many persons refuse to speak in the presence of others
   D. *separately* to prevent the testimony of one from coloring the testimony of the others

5. When a witness is reluctant to talk because he does not like to be involved in litigation, the MOST advisable of the following procedures for the investigator is to 5.____

   A. be as gentle as possible and interrogate in the form of casual questions and conversation
   B. attempt to give the witness a new interest or motive for testifying
   C. use a device, such as the association method, to elicit the desired information
   D. proceed with sternness and determination, warning the witness of the serious consequences of his refusal

6. If a person you are interviewing in connection with a character investigation obviously is not telling the truth, the MOST advisable of the following procedures is to 6.____

   A. let him talk as much as he likes so that he may eventually contradict himself and tell the truth
   B. threaten him with criminal prosecution if he does not tell the truth
   C. administer an oath to him before he is questioned
   D. disregard his testimony entirely and question him no further

7. In the course of a routine investigation of sales tax payments, the examination of a firm's books discloses to the investigator evidence that the firm's bookkeeper may be appropriating large sums of the firm's funds to his own use. 7.____
The investigator's BEST course of action, among the following, would be to

   A. warn the bookkeeper of his discovery but take no further action since his obligations are toward the city not the firm
   B. advise the firm of his suspicions, suggesting an audit of the books
   C. immediately report his findings to the District Attorney
   D. take no action other than to include the evidence among the findings in his report

8. Information obtained by an investigator from a very small child should be carefully evaluated because, of the following reasons, children 8.____

   A. are less observing than adults
   B. have less retentive memories
   C. easily confuse their own experiences with those of others
   D. are apt to have been coached by adults

9. In the course of an investigation of a claim for damages for personal injuries sustained by an individual, an anonymous letter is received by the investigator accusing this individual of mistreating his wife and children. The MOST advisable of the following courses of action for the investigator to pursue is 9.____

   A. as a law enforcement officer,to report the matter to the proper authorities
   B. to place less credence in the testimony given by the individual in view of this impeachment of his character
   C. to attempt to trace the letter and inquire further into the allegations made therein before submitting his report
   D. to disregard the letter since it has no direct bearing on the matter under investigation

10. In interviewing a person, *suggestive questions* should be avoided because, among the following, 10.____

   A. the answers to leading questions are not admissible in evidence
   B. an investigator must be fair and impartial
   C. the interrogation of a witness must be formulated according to his mentality
   D. they are less apt to lead to the truth

11. Among the following, it is GENERALLY desirable to interview a person outside his home or office because 11.____

   A. the presence of relatives and friends may prevent him from speaking freely
   B. a person's surroundings tend to color his testimony

C. the person will find less distraction outside his home or office
D. a person tends to dominate the interview when in familiar surroundings.

12. Even when an investigator is convinced of the honesty and truthfulness of a witness, thorough checking of all reported information with physical facts is imperative because, among the following, 12.____

A. mere parole testimony is not accepted as legal evidence
B. the observation of the witness may have been imperfect due to some factors which distort normal sensory perception
C. the physical facts may have changed since they were observed by the witness
D. an interview with a witness is merely an informal questioning conducted to learn facts

13. If the memory of a witness fails him about the time of an occurrence concerning which he is being questioned, the MOST advisable of the following procedures for the investigator to follow is to 13.____

A. supply the data for him in his report
B. assume the presence of a motive for concealing the information
C. request him to make an affidavit to that effect
D. try to give him some associated ideas to refresh his memory

14. If a person interviewed seems hesitant to talk while the investigator is taking notes, the MOST advisable of the following procedures for the investigator is to 14.____

A. adjourn the interview until a time when it can be conducted in a place with a hidden microphone to record it
B. secure his cooperation by explaining to the witness the importance of full and complete notes for good investigation reports
C. complete the interview without notetaking and, at the first opportunity after the interview, reduce it to writing
D. administer an oath to the person so that he will commit perjury by failing to tell the whole truth

15. The personal interview as a means of obtaining information about past occurrences is 15.____

A. the most reliable and accurate method
B. useful principally as a means of finding clues to more reliable sources of information
C. generally as reliable as recourse to documentary sources
D. qualitatively inferior but quantitatively superior to all other methods

16. Experiments have shown that the MOST satisfactory method, among the following, for obtaining dependable data in an interview is by employment of 16.____

A. the free narrative method, in which the person interviewed is permitted to talk without interruption
B. the question-and-answer method, in which the person interviewed gives information only in response to questions

C. a combination of the question-and-answer and free narrative methods, with the free narrative given first
D. a combination of the question-and-answer and free narrative methods, with the question-and-answer interview given first

17. Interviewing witnesses by the question-and-answer method, rather than allowing the witness to tell his story without interruption, will GENERALLY _______ accuracy of the report. 17.____

A. *increase* the range but decrease the
B. *decrease* the range but increase the
C. *decrease* both the range and
D. *increase* both the range and

18. Among the following, the present good health of a disabled war veteran is BEST indicated by 18.____

A. a recently issued life insurance policy
B. a return to his pre-war employment as a cashier
C. his withdrawal of a civil service veteran preference claim
D. a reduction in the amount of his pension by the Veterans Administration.

19. Among the following, a person's general good character is BEST evidenced by 19.____

A. the absence of an F.B.I. record
B. a Police Department good conduct certificate
C. his school and employment records
D. letters of recommendation he obtains from friends

20. Among the following, the signature cards of a bank might be employed as a means of verifying an individual's 20.____

A. character
B. identity
C. financial status
D. employment

21. Of the following, the overt item of evidence which most strongly indicates that an adult person is probably NOT a citizen is 21.____

A. the fact he associates frequently with recently arrived aliens
B. his lack of a birth certificate
C. his inability to speak English
D. the fact his parents are aliens

22. Among the following, an original birth certificate may serve as proof of age and 22.____

A. physical condition
B. religion
C. citizenship
D. residence

23. *Prima facie* evidence is evidence which 23.____

A. suffices to establish a fact unless rebutted or until overcome by other evidence
B. has not been tested or measured as to its validity
C. shows the existence of one fact by proof of the existence of other facts from which the first may be inferred
D. results from certain presumptions of law, which may not have a basis in fact

24. A copy, accompanied by a certificate of the proceedings necessary to be taken in order to authorize the same to be entered of record, is called a(n) _______ copy. 24.____

A. exempliﬁed
B. certiﬁed
C. true
D. veriﬁed

25. The term *surveillance*, as used in connection with investigations, is synonymous with 25.____

A. undercover work
B. reconnaissance
C. shadowing
D. inspection

26. An investigation manual directs that all investigators' reports contain a precis. The term *precis* is synonymous with 26.____

A. extract
B. paraphrase
C. synopsis
D. conclusion

27. A sworn statement made by the person who served a summons, setting forth the place and manner of service, is called a(n) _______ of service. 27.____

A. admission
B. affidavit
C. certificate
D. acknowledgment

28. A book in which deeds are recorded in the City Registrar's Office is referred to as a 28.____

A. text B. folio C. volume D. liber

29. The system of describing persons that is GENERALLY employed by modern investigators is known as the _______ system. 29.____

A. Bertillion
B. Henry
C. Moulage
D. Portrait Parle

30. A satisfaction piece is an instrument 30.____

A. which purports to discharge land from the lien of a mortgage
B. by which pending litigation is settled out of court
C. acknowledging payment of a money judgment
D. by which a lien on personal property is discharged

31. The part of an instrument which reads: *Sworn to before me this eighteenth day of July, 2015, Joseph Smith, Notary Public, State of* ..... is known as the 31.____

A. jurat
B. authentication
C. certification
D. attestation

32. An authority for the arrest of a person on a criminal charge with a view to his trial and commitment thereon is called a 32.____

A. subpoena B. summons C. complaint D. warrant

33. Entries on the block index sheets for conveyances in the city Registrar's Office are made in _______ order. 33.____

A. alphabetical
B. date
C. numerical
D. no particular

34. Generally, a summons may be served 34.____

A. at any hour of the day or night any day of the week
B. between sunrise and sunset on any day of the week
C. at any hour of the day or night, any day of the week, except Sunday
D. between sunrise and sunset on any day of the week, except Sunday

35. A party refuses to accept service of a summons when properly offered him. Among the following methods, personal service upon him could be properly made after informing him of the nature of the instrument by 35.____

A. thrusting the summons into his lap or upon his person
B. sending the summons to him by registered mail
C. leaving the summons on a table before him in his presence
D. leaving the summons with another member of his household

36. In an action against the city, personal service of the summons is made by delivering a copy thereof to the mayor, 36.____

A. treasurer, or city clerk
B. comptroller or city clerk
C. treasurer or corporation counsel
D. comptroller or corporation counsel

37. Among the following, the present home and business address of a member of the board of directors of a city bank may MOST readily be obtained from 37.____

A. MOODY'S BANK AND FINANCE DIRECTORY
B. POOR'S REGISTER
C. TROW'S CITY DIRECTORY
D. DAU'S BLUE BOOK

38. Among the following, a list of the names and addresses of various professional associations and societies in the United States would be found in the 38.____

A. WORLD ALMANAC
B. ENCYCLOPEDIA BRITANNICA
C. CONGRESSIONAL RECORD
D. POOR'S REGISTER

39. Workmen's Compensation claims are filed with the 39.____

A. Comptroller
B. State Department of Labor
C. United States Department of Labor
D. Municipal Compensation Board

40. The Bureau of Narcotics is part of the United States 40.____

A. Department of Agriculture
B. Department of Commerce
C. Treasury Department
D. Justice Department

DIRECTIONS: Questions 41 through 49 are to be answered on the basis of the following passage.

Assume that in an interview John Jones, the subject of an investigation, gave the following information concerning himself:

I was born on January 5, 1948 in a hospital in Manhattan. My parents resided in Brooklyn at the time. My mother was born in this country, but my father was born in England and came to this country with his parents when he was about eighteen years old. I attended P.S. 300 in Brooklyn from 1954 to 1962 and Central High School in Brooklyn from 1962 to 1966. From 1966 to 1972, I was employed as a clerk by the XYZ Corporation, which has since gone out of business and been dissolved. In 1972, I was inducted into the Army at Fort Dix, New Jersey, from Selective Service Board No. 528, Brooklyn. I served with the 1097th Infantry Regiment until my discharge on December 1, 1975. After that, I was employed as a merchant seaman on the U.S. Barclay by the Red Circle Steamship Company.

41. Among the following, Jones' birth record should be sought in the 41.____

A. County Clerk's Office of the county in which he was born
B. City Clerk's Office
C. Department of Health
D. Register's Office

42. If no record of Jones' birth is available there and he has no baptismal record, among the following, his date of birth may BEST be verified from the records of 42.____

A. P.S. 300, Brooklyn
B. the Board of Elections
C. Draft Board No. 528, Brooklyn
D. the XYZ Corporation

43. Among the following, Jones' father's citizenship can be verified through records of the 43.____

A. Department of State
B. Immigration and Naturalization Service
C. Federal Bureau of Investigation
D. U.S. Customs Service

44. If, in verifying Jones' education, it is desirable to write to the principal of Central High School, among the following, his full name and the address of the school will be found in the 44.____

A. Brooklyn Telephone Directory
B. NEW YORK CITY OFFICIAL DIRECTORY (THE GREEN BOOK)
C. Civil List
D. DIRECTORY OF AMERICAN SCHOOLS AND COLLEGES

45. Among the following, data regarding the now defunct XYZ Corporation could BEST be obtained from the records of the 45.____

A. New York City Department of Commerce
B. United States Department of Commerce
C. Secretary of State of New York State
D. Attorney General of New York State

46. Among the following, an inquiry regarding Jones' military service should be addressed to the 46.____

A. Chairman, Selective Service Board No. 528, Brooklyn, N.Y.
B. Adjutant General, The Pentagon, Washington, D.C.
C. Commanding Officer, 1097th Infantry Regiment, Defense Department, Washington, D.C.
D. Regional Office, Veterans Administration, Brooklyn, N.Y.

47. Among the following, if none of the officers of the XYZ Corporation can be located, Jones' employment with that Corporation might BEST be verified from the records of the 47.____

A. State Department of Labor
B. Secretary of State
C. Social Security Administration
D. Superintendent of Insurance

48. Among the following, data regarding Jones' financial and credit status can BEST be obtained from 48.____

A. R.L. POLK DIRECTORY
B. DUN AND BRADSTREET'S
C. Municipal Credit Union
D. Tax Department

49. If Jones did not want his present employer to know of the pending investigation, from among the following, his employment on the S.S. Barclay could be verified from the records of the 49.____

A. U.S. Maritime Commission
B. Bureau of Customs
C. Port of New York Authority
D. Department of Marine and Aviation

Questions 50-57.

DIRECTIONS: Column I lists various records or instruments. Column II lists various public offices. In the space at the right, opposite the number preceding each of the records or instruments in Column I, place the letter preceding the public office in Column II in which such record or information concerning such instrument may be obtained.

COLUMN I

50. Voting records
51. Record of deaths
52. Record of marriages performed in 2015
53. Record of marriages performed in 1953
54. Birth certificates
55. Lis pendens in a real property action
56. Will relating to real property
57. Certificate of appointment of a notary public

COLUMN II

A. County Clerk's Office
B. Surrogate's Office
C. Board of Elections
D. Department of Health

Questions 58-66.

DIRECTIONS: Column I lists various records. Column II lists various governmental departments and offices. In the space at the right, opposite the number preceding each of the records in Column I, place the letter preceding the department or office in Column II from which you would seek information regarding such record.

COLUMN I

58. Real estate tax assessment rolls
59. Compensating use tax records
60. Personal injury claims against the city
61. Licenses as Commissioner of Deed
62. Pawnbroker's license records
63. Pistol license records
64. Zoning regulations
65. Real estate mortgage records
66. Public assistance records

COLUMN II

A. Health Department
B. Comptroller's Office
C. City Register's Office
D. Department of Finance
E. Department of Licenses
F. Police Department
G. City Marshal
H. Tax Department
I. Bureau of Real Estate
J. City Clerk's Office
K. City Planning Commission
L. Welfare Department

Questions 67-74.

DIRECTIONS: Column I lists various licenses and records. Column II lists various state departments and offices. In the space at the right, opposite the number preceding each of the licenses or records in Column I, place the letter preceding the department or office in Column II from which you would seek information regarding such license or record.

COLUMN I

67. Bailbondsman's license
68. Retail liquor store permit
69. Certificate of incorporation of stock corporation
70. Real estate broker's license
71. Physician's license
72. Income tax records
73. License as private investigator
74. State criminal identification records

COLUMN II

A. Secretary of State
B. Department of Education
C. Department of Taxation and Finance
D. Department of Audit and Control
E. Alcoholic Beverage Control Division
F. Division of Parole
G. Department of Correction H. Department of Labor
H. Banking Department J. Insurance Department
I. Department of Social Services

Questions 75-80.

DIRECTIONS: Column I lists various records or instruments. Column II lists various federal departments or offices. In the space at the right, opposite the number preceding each of the records or instruments in Column I, place the letter preceding the office or department in Column II from which information regarding such record or instrument may be obtained.

COLUMN I

75. Passport records
76. Register of copyrights
77. Federal income tax records
78. Vietnam War draft records
79. Immigration visas
80. Bankruptcy petitions

COLUMN II

A. Department of State
B. Library of Congress
C. U.S. District Court
D. Social Security Administration
E. U.S. Patent Office
F. Office of Selective Service Records
G. Internal Revenue Bureau
H. Adjutant General's Office
I. Department of Justice

KEY (CORRECT ANSWERS)

1.	B	21.	C	41.	C	61.	J
2.	A	22.	C	42.	A	62.	E
3.	D	23.	A	43.	B	63.	F
4.	D	24.	A	44.	B	64.	K
5.	B	25.	C	45.	C	65.	C
6.	A	26.	C	46.	B	66.	L
7.	B	27.	B	47.	C	67.	J
8.	C	28.	D	48.	B	68.	E
9.	D	29.	D	49.	A	69.	A
10.	D	30.	A	50.	C	70.	A
11.	A	31.	A	51.	D	71.	B
12.	B	32.	D	52.	A	72.	C
13.	D	33.	B	53.	D	73.	A
14.	C	34.	C	54.	D	74.	G
15.	B	35.	C	55.	A	75.	A
16.	C	36.	D	56.	B	76.	B
17.	A	37.	B	57.	A	77.	G
18.	A	38.	A	58.	H	78.	F
19.	C	39.	B	59.	B	79.	A
20.	B	40.	D	60.	B	80.	C

EXAMINATION SECTION

TEST 1

DIRECTIONS: Each question or incomplete statement is followed by several suggested answers or completions. Select the one that BEST answers the question or completes the statement. *PRINT THE LETTER OF THE CORRECT ANSWER IN THE SPACE AT THE RIGHT.*

1. An investigator uses Forms A, B, and C in filling out his investigation reports. He uses Form B five times as often as Form A, and he uses Form C three times as often as Form B.
If the total number of all forms used by the investigator in a month equal 735, how many times was Form B used? 1.____
A. 150 B. 175 C. 205 D. 235

2. Of all the investigators in one agency, 25% work in a particular building. Of these, 12% have desks on the 14th floor.
What percentage of the investigators work in this building but do NOT have desks on the 14th floor? 2.____
A. 12% B. 13% C. 22% D. 23%

3. An investigator is given two reports to read. Report P is 160 pages long and takes the investigator 3 hours and 20 minutes to read.
If Report S is 254 pages long and the investigator reads it at the same rate as he reads Report P, how long will it take him to read Report S? _____ hours _____ minutes. 3.____
A. 4; 15 B. 4; 50 C. 5; 10 D. 5; 30

4. A team of 6 investigators was assigned to interview 234 people.
If half the investigators conduct twice as many interviews as the other half, and the slow group interviews 12 persons a day, how many days would it take to complete this assignment? _____ days. 4.____
A. 4½ B. 5 C. 6 D. 6½

5. The investigators in one agency conduct an average of 12 interviews an hour from 10 A.M. to 12 noon and from 1 P.M. to 5 P.M. daily. The director of his agency knows from past experience that 20% of those called in to be interviewed are unable to keep the appointments that were scheduled.
If the director wants his staff to be kept occupied with interviews for the entire time period that has been set aside for this function, how many appointments should be scheduled for each day? 5.____
A. 86 B. 90 C. 96 D. 101

6. An investigator has a 430-page report to read. The first day, he is able to read 20 pages. The second day, he reads 10 pages more than the first day, and the third day, he reads 15 pages more than the second day. 6.____

If, on the following days, he continues to read at the same rate as he was reading on the third day, he will complete the report on the _____ day.
A. 7th B. 8th C. 10th D. 11th

7. The 36 investigators in an agency are each required to submit 25 investigation reports a week. These reports are filled out on a certain form, and only one copy of the form is needed per report. 7.____
Allowing 20% for waste, how many packages of 45 forms a piece should be ordered for each weekly period?
A. 15 B. 20 C. 25 D. 30

8. During the fiscal year, an investigative unit received $260 for stationery and telephone expenditures. It spent 43% for stationery and 1/3 of the balance for telephone service. 8.____
The amount of money that was left at the end of the fiscal year was MOST NEARLY
A. $49 B. $50 C. $99 D. $109

Questions 9-10.

DIRECTIONS: Questions 9 and 10 are to be answered SOLELY on the data given below.

Number of days absent per worker (sickness)	1	2	3	4	5	6	7	8 or Over
Number of Workers	96	45	16	3	1	0	1	0

Total Number of Workers: 500

9. The TOTAL number of man days lost due to illness in 2020 was 9.____
A. 137 B. 154 C. 162 D. 258

10. Of the 500 workers studied, the number who lost NO days due to sickness in 2020 was 10.____
A. 230 B. 298 C. 338 D. 372

Questions 11-13.

DIRECTIONS: Questions 11 through 13 are to be answered SOLELY on the basis of the following passage.

The rise of urban-industrial society has complicated the social arrangements needed to regulate contacts between people. As a consequence, there has been an unprecedented increase in the volume of laws and regulations designed to control individual conduct and to govern the relationship of the individual to others. In a century, there has been an eight-fold increase in the crimes for which one may be prosecuted.

For these offenses, the courts have the ultimate responsibility for redressing wrongs and convicting the guilty. The body of legal precepts gives the impression of an abstract and evenhanded dispensation of justice. Actually, the personnel of the agencies applying these precepts are faced with the difficulties of fitting abstract principles to highly variable situations emerging from the dynamics of everyday life. It is inevitable that discrepancies should exist between precept and practice.

The legal institutions serve as a framework for the social order by their slowness to respond to the caprices of transitory fad. This valuable contribution exacts a price in terms of the inflexibility of legal institutions in responding to new circumstances. This possibility is promoted by the changes in values and norms of the dynamic larger culture of which the legal precepts are a part.

11. According to the above passage, the increase in the number of laws and regulations during the twentieth century can be attributed to the 11.____
   A. complexity of modern industrial society
   B. increased seriousness of offenses committed
   C. growth of individualism
   D. anonymity of urban living

12. According to the above passage, which of the following presents a problem to the staff of legal agencies? The 12.____
   A. need to eliminate the discrepancy between precept and practice
   B. necessity to apply abstract legal precepts to rapidly changing conditions
   C. responsibility for reducing the number of abstract legal principles
   D. responsibility for understanding offenses in terms of the real-life situations from which they emerge

13. According to the above passage, it can be concluded that legal institutions affect social institutions by 13.____
   A. preventing change
   B. keeping pace with its norms and values
   C. changing its norms and values
   D. providing stability

Questions 14-16.

DIRECTIONS: Questions 14 through 16 are to be answered SOLELY on the basis of information given in the following passage.

A personnel interviewer, selecting job applicants, may find that he reacts badly to some people even on first contact. This reaction cannot usually be explained by things that the interviewee has done or said. Most of us have had the experience of liking or disliking, of feeling comfortable and uncomfortable with people on first acquaintance, long before we have had a chance to make a conscious, rational decision about them. Often, too, our liking or disliking is transmitted to the other person by subtle processes such as gestures, posture, voice intonations, or choice of words. The point to be kept in mind is this: the relations between people are complex and occur at several levels, from the conscious to the unconscious. This is true whether the relationship is brief or long, formal or informal.

Some of the major dynamics of personality which operate on the unconscious level are projection, sublimation, rationalization, and repression. Encountering these for the first time, one is apt to think of them as representing pathological states. In the extreme, they undoubtedly are, but they exist so universally that we must consider them also to be parts of normal personality.

Without necessarily subscribing to any of the numerous theories of personality, it is possible to describe personality in terms of certain important aspects or elements. We are all aware of ourselves as thinking organisms.

This aspect of personality, the conscious part, is important for understanding human behavior, but it is not enough. Many find it hard to accept the notion that each person also has an unconscious. The existence of the unconscious is no longer a matter of debate. It is not possible to estimate at all precisely what proportion of our total psychological life is conscious, what proportion unconscious. Everyone who has studied the problem, however, agrees that consciousness is the smaller part of personality. Most of what we are and do is a result of unconscious processes. To ignore this is to risk mistakes.

14. The above passage suggests that an interviewer can be MOST effective if he 14.____
    A. learns how to determine other peoples' unconscious motivations
    B. learns how to repress his own unconsciously motivated mannerisms and behavior
    C. can keep others from feeling that he either likes or dislikes them
    D. gains an understanding of how the unconscious operates in himself and in others

15. It may be inferred from the above passage that the *subtle processes*, such as gestures, posture, voice intonation, or choice of words referred to in the first paragraph are USUALLY 15.____
    A. in the complete control of an expert investigator
    B. the determining factors in the friendships a person establishes
    C. controlled by a person's unconscious
    D. not capable of being consciously controlled

16. The above passage implies that various different personality theories are USUALLY 16.____
    A. so numerous and different as to be valueless to an investigator
    B. in basic agreement about the importance of the unconscious
    C. understood by the investigator who strives to be effective
    D. in agreement that personality factors such as projection and repression are pathological

Questions 17-19.

DIRECTIONS: Questions 17 through 19 are to be answered SOLELY on the basis of information contained in the following passage.

No matter how well the interrogator adjusts himself to the witness and how precisely he induces the witness to describe his observations, mistakes still can be made. The mistakes made by an experienced interrogator may be comparatively few, but as far as the witness is concerned, his path is full of pitfalls. Modern "witness psychology" has shown that even the most honest and trustworthy witnesses are apt to make grave mistakes in good faith. It is, therefore, necessary that the interrogator get an idea of the weak links in the testimony in order to check up on them in the event that something appears to be strange or not quite satisfactory.

Unfortunately, modern witness psychology does not yet offer any means of directly testing the credibility of testimony. It lacks precision and method, in spite of worthwhile attempts on the part of learned men. At the same time, witness psychology, through the gathering of many experience concerning the weaknesses of human testimony, has been of invaluable service. It shows clearly that only evidence of a technical nature has absolute value as proof.

Testimony may be separated into the following stages: (1) perception; (2) observation; (3) mind fixation of the observed occurrences, in which fantasy, association of ideas, and personal judgment participate; (4) expression in oral or written form, where the testimony is transferred from one witness to another or to the interrogator. Each of these stages offers innumerable possibilities for the distortion of testimony.

17. The above passage indicates that having witnesses talk to each other before testifying is a practice which is GENERALLY 17.____
   A. *desirable*, since the witnesses will be able to correct each other's errors in observation before testimony
   B. *undesirable*, since the witnesses will collaborate on one story to tell the investigator
   C. *undesirable*, since one witness may distort his testimony because of what another witness may erroneously say
   D. *desirable*, since witnesses will become aware of discrepancies in their own testimony and can point out the discrepancies to the investigator

18. According to the above passage, the one of the following which would be the MOST reliable for use as evidence would be the testimony of a 18.____
   A. handwriting expert about a signature on a forged check
   B. trained police officer about the identity of a criminal
   C. laboratory technician about an accident he has observed
   D. psychologist who has interviewed any witness who relate conflicting stories

19. Concerning the validity of evidence, it is clear from the above passage that 19.____
   A. only evidence of a technical nature is at all valuable
   B. the testimony of witnesses is so flawed that it is usually valueless
   C. an investigator, by knowing modern witness psychology, will usually be able to perceive mistaken testimony
   D. an investigator ought to expect mistakes in even the most reliable witness testimony

Questions 20-21.

DIRECTIONS: Questions 20 and 21 are to be answered SOLELY on the basis of information given in the following passage.

Since we generally assure informants that what they say is confidential, we are not free to tell one informant what the other has told us. Even if the informant says, "*I don't care who knows it; tell anybody you want to,*" we find it wise to treat the interview as confidential. An interviewer who relates to some informants what other informants have told him is likely to stir up anxiety and suspicion. Of course, the interviewer may be able to tell an informant what he has heard without revealing the source of his information. This may be perfectly appropriate where a story has wide currency so that an informant cannot infer the source of the information. But if an event is not widely known, the mere mention of it may reveal to one informant what another informant has said about the situation. How can the data be cross-checked in these circumstances.

20. The above passage IMPLIES that the anxiety and suspicion an interviewer may arouse by telling what has been learned in other interviews is due to the 20.____
   A. lack of trust the person interviewed may have in the interviewer's honesty
   B. troublesome nature of the material which the interviewer has learned in other interviews
   C. fact that the person interviewed may not believe that permission was given to repeat the information
   D. fear of the person interviewed that what he is telling the interviewer will be repeated

21. The above passage is MOST likely part of a longer passage dealing with 21.____
   A. ways to verify data gathered in interviews
   B. the various anxieties a person being interviewed may feel
   C. the notion that people sometimes say things they do not mean
   D. ways an interview can avoid seeming suspicious

Questions 22-23.

DIRECTIONS: Questions 22 and 23 are to be answered SOLELY on the basis of information given below.

The ability to interview rests not on any single trait, but on a vast complex of them. Habits, skills, techniques, and attitudes are all involved. Competence in interviewing is acquired only after careful and diligent study, prolonged practice (preferably under supervision), and a good bit of trial and error; for interviewing is not an exact science; it is an art. Like many other arts, however, it can and must draw on science in several of its aspects.

There is always a place for individual initiative, for imaginative innovations, and for new combinations of old approaches. The skilled interviewer cannot be bound by a set of rules. Likewise, there is not a set of rules which can guarantee to the novice that his interviewing will be successful. There are, however, some accepted, general guideposts which may help the beginner to avoid mistakes, learn how to conserve this efforts, and establish effective working relationships with interviewees; to accomplish, in short, what he sets out to do.

22. According to the above passage, rules and standard techniques for interviewing are 22.____
   A. helpful for the beginner, but useless for the experienced, innovative interviewer
   B. destructive of the innovation and initiative needed for a good interviewer
   C. useful for even the experienced interviewer who may, however, sometimes go beyond them
   D. the means by which nearly anybody can become an effective interviewer

23. According to the above passage, the one of the following which is a prerequisite to competent interviewing is 23.____
   A. avoid mistakes
   B. study and practice
   C. imaginative innovation
   D. natural aptitude

Questions 24-27.

DIRECTIONS: Questions 24 through 27 are to be answered SOLELY on the basis of information given in the following passage.

The question of what material is relevant is not as simple as it might seem. Frequently, material which seems irrelevant to the inexperienced has, because of the common tendency to disguise and distort and misplace one's feelings, considerable significance. It may be necessary to let the client "ramble on" for a while in order to clear the decks, as it were, so that he may get down to things that really are on his mind. On the other hand, with an already disturbed person, it may be important for the interviewer to know when to discourage further elaboration of upsetting material. This is especially the case where the worker would be unable to do anything about it. An inexperienced interviewer might, for instance, be intrigued with the bizarre elaboration of material that the psychotic produces, but further elaboration of this might encourage the client in his instability. A too random discussion may indicate that the interviewee is not certain in what areas the interviewer is prepared to help him, and he may be seeking some direction. Or again, satisfying though it may be for the interviewer to have the interviewee tell him intimate details, such revelations sometimes need to be checked or encouraged only in small doses. An interviewee who has "talked too much" often reveals subsequent anxiety. This is illustrated by the fact that frequently after a "confessional" interview, the interviewee surprises the interviewer by being withdrawn, inarticulate, or hostile, or by breaking the next appointment.

24. Sometimes a client may reveal certain personal information to an interviewer and subsequently may feel anxious about this revelation. 24.____
   If, during an interview, a client begins to discuss very personal matters, it would be BEST to
   A. tell the client, in no uncertain terms, that you're not interested in personal details
   B. ignore the client at this point
   C. encourage the client to elaborate further on the details
   D. inform the client that the information seems to be very personal

25. The author indicates that clients with severe psychological disturbances pose an especially difficult problem for the inexperienced interviewer. 25.____
The difficulty lies in the possibility of the client
   A. becoming physically violent and harming the interviewer
   B. rambling on for a while
   C. revealing irrelevant details which may be followed by cancelled appointments
   D. reverting to an unstable state as a result of interview material

26. An interviewer should be constantly alert to the possibility of obtaining clues from the client as to the problem areas. 26.____
According to the above passage, a client who discusses topics at random may be
   A. unsure of what problems the interviewer can provide help with
   B. reluctant to discuss intimate details
   C. trying to impress the interviewer with his knowledge
   D. deciding what relevant material to elaborate on

27. The evaluation of a client's responses may reveal substantial information that may aid the interviewer in assessing the problem areas that are of concern to the client. Responses that seemed irrelevant at the time of the interview may be of significance because 27.____
   A. considerable significance is attached to all relevant material
   B. emotional feelings are frequently masked
   C. an initial *rambling on* is often a prelude to what is actually bothering the client
   D. disturbed clients often reveal subsequent anxiety

Questions 28-30.

DIRECTIONS: Questions 28 through 30 are to be answered SOLELY on the basis of the following passage.

The physical setting of the interview may determine its entire potentiality. Some degree of privacy and a comfortable relaxed atmosphere are important. The interviewee is not encouraged to give much more than his name and address if the interviewer seems busy with other things, if people are rushing about, if there are distracting noises. He has a right to feel that, whether the interview lasts five minutes or an hour, he has, for that time, the undivided attention of the interviewer. Interruptions, telephone calls, and so on, should be reduced to a minimum. If the interviewee has waited in a crowded room for what seems to him an interminably long period, he is naturally in mood to sit down and discuss what is on his mind. Indeed, by that time, the primary thing on his mind may be his irritation at being kept waiting, and he frequently feels it would be impolite to express this. If a wait or interruptions have been unavoidable, it is always helpful to give the client some recognition that these are disturbing and that we can naturally understand that they make it more difficult for him to proceed. At the same time, if he protests that they have not troubled him, the interviewer can best accept his statements at their face value, as further insistence that they must have been disturbing may be interpreted by him as accusing, and he may conclude that the interviewer has been personally hurt by his irritation.

28. Distraction during an interview may tend to limit the client's responses. In a case where an interruption has occurred, it would be BEST for the investigator to 28.____
    A. terminate this interview and have it rescheduled for another time period
    B. ignore the interruption since it is not continuous
    C. express his understanding that the distraction can cause the client to feel disturbed
    D. accept the client's protests that he has been troubled by the interruption

29. To maximize the rapport that can be established with the client, an appropriate physical setting is necessary. At the very least, some privacy would be necessary. In addition, the interviewer should 29.____
    A. always appear to be busy in order to impress the client
    B. focus his attention only on the client
    C. accept all the client's statements as being valid
    D. stress the importance of the interview to the client

30. Clients who have been waiting quite some time for their interview may, justifiably, become upset. However, a client may initially attempt to mask these feelings because he may 30.____
    A. personally hurt the interviewer
    B. want to be civil
    C. feel that the wait was unavoidable
    D. fear the consequences of his statement

KEY (CORRECT ANSWERS)

1.	B	11.	A	21.	A
2.	C	12.	B	22.	C
3.	D	13.	D	23.	B
4.	D	14.	D	24.	D
5.	B	15.	C	25.	D
6.	D	16.	B	26.	A
7.	C	17.	C	27.	B
8.	C	18.	A	28.	C
9.	D	19.	D	29.	B
10.	C	20.	D	30.	B

TEST 2

DIRECTIONS: Each question or incomplete statement is followed by several suggested answers or completions. Select the one that BEST answers the question or completes the statement. *PRINT THE LETTER OF THE CORRECT ANSWER IN THE SPACE AT THE RIGHT.*

Questions 1-5.

DIRECTIONS: In Questions 1 through 5, choose the statement which is BEST from the point of view of English usage suitable for a business report.

1. A. The client's receiving of public assistance checks at two different addresses were disclosed by the investigation. 1.____
   B. The investigation disclosed that the client was receiving public assistance checks at two different addresses.
   C. The client was found out by the investigator to be receiving public assistance checks at two different addresses.
   D. The client has been receiving public assistance checks at two different addresses, disclosed the investigation

2. A. The investigation of complaints are usually handled by this unit, which deals with internal security problems in the department. 2.____
   B. This unit deals with internal security problems in the department; usually investigating complaints.
   C. Investigating complaints is this unit's job, being that it handles internal security problems in the department
   D. This unit deals with internal security problems in the department and usually investigates complaints.

3. A. The delay in completing this investigation was caused by difficulty in obtaining the required documents from the candidate. 3.____
   B. Because of difficulty in obtaining the required documents from the candidate is the reason that there was a delay in completing this investigation.
   C. Having had difficulty in obtaining the required documents from the candidate, there was a delay in completing this investigation.
   D. Difficulty in obtaining the required documents from the candidate had the affect of delaying the completion of this investigation.

4. A. This report, together with documents supporting our recommendation, are being submitted for your approval. 4.____
   B. Documents supporting our recommendation is being submitted with the report for your approval.
   C. This report, together with documents supporting our documentation, is being submitted for your approval.
   D. The report and documents supporting our recommendation is being submitted for your approval.

5. A. Several people were interviewed and numerous letters were sent before this case was completed. 5.____
   B. Completing this case, interviewing several people and sending numerous letters were necessary.
   C. To complete this case needed interviewing several people and sending numerous letters.
   D. Interviewing several people and sending numerous letters was necessary to complete the case.

Questions 6-20.

DIRECTIONS: For each of the sentences numbered 6 to 20, select from the options given below the MOST applicable choice, and mark your answer accordingly.
A. The sentence is correct.
B. The sentence contains a spelling error only.
C. The sentence contains an English grammar error only.
D. The sentence contains both a spelling error and an English grammar error.

6. He is a very dependible person whom we expect will be an asset to this division. 6.____

7. An investigator often finds it necessary to be very diplomatic when conducting an interview. 7.____

8. Accurate detail is especially important if court action results from an investigation. 8.____

9. The report was signed by him and I since we conducted the investigation jointly. 9.____

10. Upon receipt of the complaint, an inquiry was begun. 10.____

11. An employee has to organize his time so that he can handle his workload efficiantly. 11.____

12. It was not apparent that anyone was living at the address given by the client. 12.____

13. According to regulations, there is to be at least three attempts made to locate the client. 13.____

14. Neither the inmate nor the correction officer was willing to sign a formal statement. 14.____

15. It is our opinion that one of the persons interviewed were lying. 15.____

16. We interviewed both clients and departmental personel in the course of this investigation. 16.____

17. It is concievable that further research might produce additional evidence. 17.____

18. There are too many occurences of this nature to ignore. 18.____

19. We cannot accede to the candidate's request. 19.____

20. The submission of overdue reports is the reason that there was a delay in completion of this investigation. 20.____

Questions 21-2.

DIRECTIONS: Each of Questions 21 through 25 consists of three sentences lettered A, B, and C. In each of these questions, one of the sentences may contain an error in grammar, sentence structure, or punctuation, or all three sentences may be correct. If one of the sentences in a question contains an error in grammar, sentence structure, or punctuation, print in the space at the right the capital letter preceding the sentence which contains the error. If all three sentences are correct, print the letter D.

21. A. Mr. Smith appears to be less competent than I in performing these duties. 21.____
    B. The supervisor spoke to the employee, who had made the error, but did not reprimand him.
    C. When he found the book lying on the table, he immediately notified the owner.

22. A. Being locked in the desk, we were certain that the papers would not be taken. 22.____
    B. It wasn't I who dictated the telegram; I believe it was Eleanor.
    C. You should interview whoever comes to the office today.

23. A. The clerk was instructed to set the machine on the table before summoning the manager. 23.____
    B. He said that he was not familiar with those kind of activities.
    C. A box of pencils, in addition to erasers and blotters, was included in the shipment.

24. A. The supervisor remarked, "Assigning an employee to the proper type of work is not always easy." 24.____
    B. The employer found that each of the applicants were qualified to perform the duties of the position.
    C. Any competent student is permitted to take this course if he obtains the consent of the instructor.

25. A. The prize was awarded to the employee whom the judges believed to be most deserving. 25.____
    B. Since the instructor believes this book is the better of the two, he is recommending it for use in the school.
    C. It was obvious to the employees that the completion of the task by the scheduled date would require their working overtime.

KEY (CORRECT ANSWERS)

1. B
2. D
3. A
4. C
5. A

6. D
7. A
8. A
9. C
10. A

11. B
12. B
13. C
14. A
15. C

16. B
17. B
18. B
19. A
20. C

21. B
22. A
23. B
24. B
25. D

EXAMINATION SECTION
TEST 1

DIRECTIONS: Each question or incomplete statement is followed by several suggested answers or completions. Select the one that BEST answers the question or completes the statement. *PRINT THE LETTER OF THE CORRECT ANSWER IN THE SPACE AT THE RIGHT.*

1. One of your men is doing a new job incorrectly. 1.____
The BEST action for you to take is to
   A. criticize him in the presence of the other men
   B. criticize him in private
   C. bring him up on charges
   D. show him how to do it correctly

2. Of the following, the BEST reason why it is unacceptable policy for you to become too friendly with the men you supervise is that the men may 2.____
   A. try to take advantage of your friendship
   B. resent your familiarity
   C. wish to borrow money from you
   D. be transferred to another unit

3. Of the following, the attitude for you to have toward your men in order to accomplish your job BEST is to be 3.____
   A. harsh and uncompromising
   B. firm and fair
   C. easygoing and forgiving
   D. aloof and unsocial

4. A man in your gang complains that the work is dirty. 4.____
Of the following, the BEST action for you to take is to
   A. give the man only the clean jobs
   B. tell the man that the dirt is part of the working conditions
   C. tell the man to quit if he does not like the working conditions
   D. bring the man up on charges

5. Although you estimate that you will need 4 men to do a certain job, you bring 6 men to do the job. 5.____
This practice is considered by authorities to be
   A. *good,* since you will be sure to get the job done on time
   B. *good,* since some men may get sick on the job and may be unable to work
   C. *poor,* since men may stand around doing nothing
   D. *poor,* since the work will not be divided evenly

6. One of your men tends to *goof off* whenever he has the chance. 6.____
Of the following, the BEST procedure to follow first with respect to this man is to
   A. have him transferred to another unit
   B. deduct the estimated wasted time from his time off
   C. give him the hardest jobs
   D. watch him closely

7. If four men work seven hours during the day, the number of man-hours of work done is 7.____

A. 4 B. 7 C. 11 D. 28

8. You should check that you have all the equipment and material you need for the day before work is started. Of the following, the BEST reason for personally making this check is that 8.____

A. the men under your supervision cannot be trusted
B. the men are usually too busy to check the material and equipment
C. it is your responsibility to see that everything is in order
D. it is very difficult to get help for checking once you are in the field

9. One of your men is injured on the job.
The FIRST thing you should do is to 9.____

A. assist the injured man
B. find out the circumstances of the accident
C. call the office to notify your supervisor of the accident
D. fill out the paperwork relating to the accident

10. When investigating a complaint by a home owner of sewage backing up in a house, you find that the house trap in the basement is blocked.
Of the following, the PROPER action for you to take is to 10.____

A. call in a plumber for the home owner
B. clean out the house trap
C. tell the home owner to call in a plumber
D. disconnect the house trap from the piping, clean it out, and reinstall the trap

11. If it takes four men fourteen days to do a certain job, seven men, working at the same rate, should be able to do the same job in ______ days. 11.____

A. 8 B. 7 C. 6 D. 5

12. The men you supervise suggest that work be started an hour earlier so that they can leave an hour earlier at the end of the day.
Of the following, the BEST action for you to take is to 12.____

A. ignore the request
B. start work an hour earlier
C. tell them you will forward their suggestion to your superior
D. report the men for insubordination

13. One of the men under your supervision tells you he is ill and would like to leave the job.
Of the following, the BEST action for you to take is to 13.____

A. grant the request
B. report the man for trying to goof off
C. take the man personally to the department doctor
D. tell the man he has to work the rest of the day or he will lose a day's pay

14. One of your men scheduled to arrive at 8 A.M. calls you at noon to inform you that he will not be in because of personal business. 14.____
Of the following, the BEST action for you to take FIRST is to

A. tell him to take it off sick leave
B. call the office and ask for a replacement
C. tell the man he should have called in on or about 8 A.M.
D. tell him to charge the absence to lateness

15. Assume that you are in the field and have completed your work 2 hours before quitting time. The men spend the remaining 2 hours sitting in a restaurant. 15.____
This practice is considered by authorities to be

A. *good,* as the men put in a full day
B. *good,* as make–work is a poor policy
C. *poor,* because it creates a bad public image
D. *poor,* as it disrupts the restaurant's business

16. As a foreman, you insist that all mechanical equipment you use be PROPERLY serviced and maintained by your men. This policy is 16.____

A. *poor,* since you may be pressuring the men
B. *poor,* since the men may not cooperate
C. *good,* since it helps prevent breakdown in equipment which can cause work to stop
D. *good,* since the equipment is serviced on the men's time so that you get more work out of the men

17. Assume that the men you supervise are cleaning out a catchbasin and uncover a gun. 17.____
Of the following, the BEST action to take is to

A. notify the police department of the discovery
B. throw the gun away because it probably does not work
C. keep the gun since you may be able to repair it
D. dismantle the gun before disposing of it because it may be loaded

18. While your crew is working, a passer–by stops and asks you what they are doing. 18.____
Of the following, the BEST action to take is to

A. tell him to mind his own business
B. briefly explain your operation
C. tell him to write a letter to the sewer department
D. ignore the man and call the police if he persists

19. Your men should be careful not to break manhole covers. Of the following, the BEST reason for taking this precaution is that 19.____

A. the cost of the manhole cover will be taken out of your paycheck
B. the manhole cover can't be replaced
C. manhole covers cost money to replace
D. broken manhole covers are difficult to get rid of

20. While on the job, you teach your duties to one of the laborers. This practice is considered by authorities to be 20.____
    A. *poor,* because it shows favoritism
    B. *poor,* because this laborer may undermine your authority
    C. *good,* because the laborer will then be able to pass a promotion examination
    D. *good,* because the laborer can replace you in an emergency

KEY (CORRECT ANSWERS)

1.	D	11.	A
2.	A	12.	C
3.	B	13.	A
4.	B	14.	C
5.	C	15.	C
6.	D	16.	C
7.	D	17.	A
8.	C	18.	B
9.	A	19.	C
10.	C	20.	D

TEST 2

DIRECTIONS: Each question or incomplete statement is followed by several suggested answers or completions. Select the one that BEST answers the question or completes the statement. *PRINT THE LETTER OF THE CORRECT ANSWER IN THE SPACE AT THE RIGHT.*

1. The BEST reason for you to advise your men to be alert at all times while working in the street is that 1.____

   A. working in the street could be dangerous
   B. they may see some criminal activity
   C. somebody from the main office may be observing your men
   D. they may create a bad public image if they are not always alert

2. It is GOOD practice to complete a report on an accident as soon as possible after the accident occurs MAINLY because 2.____

   A. paperwork should be submitted to the office on the same day an accident occurs
   B. if you do not you may forget some of the necessary details
   C. this gives you more time to change the report if this should be necessary
   D. the department can then immediately prepare its defense

3. Official directives state that you are to report immediately by telephone if a manhole cover or basin grate is missing. 3.____
   Of the following, the BEST reason for having this requirement is to

   A. permit the cover or grate to be ordered if it is not on hand
   B. be able to assess the responsibility for this condition
   C. prevent an accident
   D. enable the sanitation department to clean the street

4. A complainant is a 4.____

   A. city agency that responds to a complaint
   B. person filing a complaint
   C. crew member that responds to a complaint
   D. lawyer who defends a client against a complaint

5. In filling out an accident form, there is a section entitled *Accident Type.* 5.____
   Of the following, the one that is an accident type is

   A. struck by falling object
   B. operated without authority
   C. worked too slowly
   D. engaged in horseplay

6. On an accident report, there is an item labeled *Nature of Injury.* 6.____
   Of the following, the one that belongs in this category is

   A. fracture B. carelessness
   C. defective equipment D. loose clothing

7. Of the following, the LEAST serious of the defects filed in a sewer report is 7.____

   A. broken casting B. missing casting
   C. noisy manhole cover D. backed up sewer

8. When signing a time sheet, the employee must sign his name and his number. 8._____
The BEST of the following reasons for requiring his number in addition to his name is

A. to be sure the employee has not entered the wrong time on the time sheet
B. to make it easier to contact the employee
C. his signature may be difficult to read
D. the employee is paid based on his number which is fed into the IBM machine

9. One of the men in your unit states that he will take off the next day to attend his father–in–law's funeral and wants to know if he can change the absence to sick leave. Of the following, the BEST answer you can give him is that 9._____

A. he can charge half the time to sick leave and half to annual leave
B. the rules do not permit this to be done
C. this can only be done if his father–in–law had lived with him
D. sick leave can be used this way only if he had 10 years or more in service

10. In addition to the Department of Water Resources, the Environmental Protection Administration consists of the 10._____

A. Board of Water Supply, the Department of Sanitation, and the Department of Air Resources
B. Department of Sanitation, the Department of Municipal Services, and the Department of Air Resources
C. Department of Sanitation and the Department of Municipal Services
D. Department of Sanitation and the Department of Air Resources

11. The government calendar year starts on ______ 1. 11._____

A. June B. July C. May D. January

12. A truck leaves the garage at 9:26 A.M. and returns the same day at 3:43 P.M. 12._____
The period of time that the truck was away from the garage is MOST NEARLY ______ hours ______ minutes.

A. 5; 17 B. 5; 43 C. 6; 17 D. 6; 26

13. Of the following, the BEST method for a foreman to use to teach a man how to lift a manhole cover safely is to 13._____

A. tell him how to do it
B. make a sketch showing the correct method to use
C. actually lift a cover with the man watching
D. let the man try to lift the cover and correct any mistakes

14. Assume that one of the laborers you supervise is unable to read well and that you have advised him to go evenings to school to learn to read and write English. 14._____
According to good supervisory practice, the advice is considered to be

A. *poor,* because it is none of your business
B. *poor,* because a laborer does not have to know how to read
C. *good,* because he can then go on to get a high school diploma
D. *good,* because he will be able to read signs and avoid danger on the job

15. Assume that a new piece of mechanical equipment is brought to the job. 15._____
Of the following, the BEST way for the men to learn the proper use of the equipment is to

A. have a representative of the company that manufactures the equipment come to the job and demonstrate its use
B. let the men try out the equipment and learn the operation of the equipment by using it
C. let the men read the instruction manual carefully before trying out the equipment
D. deliver a lecture to the men that have to use the equipment on the proper use of the equipment

16. Assume that you are training a group of men on the adjustment of a high–pressure relief valve. 16._____
Of the following, the FIRST topic you should discuss with the men is

A. the conditions under which it is necessary to adjust the relief valve
B. how to order parts for the relief valve
C. how the springs in the relief valve work
D. how to take apart the relief valve

17. Assume that a new man is assigned to your unit and you explain to him exactly what is expected of him. 17._____
This procedure is

A. *poor,* because the new man will feel that you are threatening him
B. *poor,* because this leaves the new man with no freedom to do the job as he feels best
C. *good,* because then the new man can quit if he does not like the foreman
D. *good,* because the new man will know what is required of him

18. A foreman explains to a man a way of doing a particular job and the man says he does not understand. 18._____
Of the following, the BEST action for the foreman to take is to

A. repeat the explanation
B. let the man remain ignorant
C. transfer the man to another unit
D. tell the man he may understand the procedure at a later time

19. A new piece of equipment is ordered and the men who will use it are trained in its use before the equipment arrives on the job. 19._____
This practice is

A. *poor,* because the order may be cancelled and time wasted
B. *poor,* because it takes longer to train men when the equipment is not present
C. *good,* because it keeps the men busy when they do not have anything to do
D. *good,* because the equipment can immediately be put to use

20. You observe a man using a piece of equipment incorrectly. Of the following, the BEST action for you to take is to 20.____

A. have somebody else work with the equipment
B. transfer the man to another unit
C. bring the man up on charges
D. show him how to use the equipment correctly

KEY (CORRECT ANSWERS)

1.	A	11.	D
2.	B	12.	C
3.	C	13.	C
4.	B	14.	D
5.	A	15.	A
6.	A	16.	A
7.	C	17.	D
8.	C	18.	A
9.	B	19.	D
10.	D	20.	D

EXAMINATION SECTION
TEST 1

DIRECTIONS: Each question or incomplete statement is followed by several suggested answers or completions. Select the one that BEST answers the question or completes the statement. *PRINT THE LETTER OF THE CORRECT ANSWER IN THE SPACE AT THE RIGHT.*

1. A *basic* method of operation that a *good* supervisor should follow is to 1.____
   A. check the work of subordinates constantly to make sure they are not making exceptions to the rules
   B. train subordinates so they can handle problems that come up regularly themselves and come to him only with special cases
   C. delegate to subordinates only those duties which he cannot do himself
   D. issue directions to subordinates only on special matters

2. To do a *good* job of performance evaluation, it is BEST for a supervisor to 2.____
   A. compare the employee's performance to that of another employee doing similar work
   B. give greatest weight to instances of unusually good or unusually poor performance
   C. leave out any consideration of the employee's personal traits
   D. measure the employee's performance against standard performance requirements

3. Of the following, the MOST important reason for a supervisor to have private face to face discussions with subordinates about their performance is to 3.____
   A. help employees improve their work
   B. give special praise to employees who perform well
   C. encourage the employees to compete for higher performance ratings
   D. discipline employees who perform poorly

4. Of the following, the CHIEF purpose of a probationary period for a new employee is to allow time for 4.____
   A. finding out whether the selection processes are satisfactory
   B. the employee to make adjustments in his home circumstances made necessary by the job
   C. the employee to decide whether he wants a permanent appointment
   D. determining the fitness of the employee to continue in the job

5. When a subordinate resigns his job, it is MOST important to conduct an exit interview in order to 5.____
   A. try to get the employee to remain on the job
   B. learn the true reasons for the employee's resignation
   C. see that the employee leaves with a good opinion of the agency
   D. ask the employee if he would consider a transfer

6. Chronic lateness of employees is generally LEAST likely to be due to 6.____
   A. distance of job location from home
   B. poor personnel administration
   C. unexpressed employee grievances
   D. low morale

7. Of the following, the LEAST effective stimulus for motivating employees toward inproved performance over a long-range period is 7.____

   A. their sense of achievement
   B. their feeling of recognition
   C. opportunity for their self-development
   D. an increase in salary

8. Suppose that NOT ONE of a group of employees has turned in an idea to the employees suggestion system during the past year.
   The *most probable* reason for this situation is that the 8.____

   A. money awards given for suggestions used are not high enough to make employees interested
   B. employees in this group are not able to develop any good ideas
   C. supervisor of these employees is not doing enough to encourage them to take part in the program
   D. methods and procedures of operation do not need improvement

9. A subordinate tells you that he is having trouble concentrating on his work due to a personal problem at home.
   Of the following, it would be BEST for you to 9.____

   A. refer him to a community service agency
   B. listen quietly to the story because he may just need a sympathetic ear
   C. tell him that you cannot help him because the problem is not job related
   D. ask him questions about the nature of the problem and tell him how you would handle it

10. For you as a supervisor to give each of your subordinates *exactly* the same type of supervision is 10.____

   A. *advisable,* because doing this insures fair and impartial treatment of each individual
   B. *not advisable,* because individuals like to think that they are receiving better treatment than others
   C. *advisable,* because once a supervisor learns how to deal with a subordinate who brings a problem to him, he can handle another subordinate with this problem in the same way
   D. *not advisable,* because each person is different and there is no one supervisory procedure for dealing with individuals that applies in every case

11. A senior employee under your supervision tells you that he is reluctant to speak to one of his subordinates about his poor work habits, because this worker is "strong-willed" and he does not want to antagonize him.
   For you to offer to speak to the subordinate about this matter yourself would be 11.____

   A. *advisable,* since you are in a position of greater authority
   B. *inadvisable,* since handling this problem is a basic supervisory responsibility of the senior employee
   C. *advisable,* since the senior employee must work more closely with the worker than you do
   D. *inadvisable,* since you should not risk antagonizing the employee yourself

12. Some of your subordinates have been coming to you with complaints you feel are unimportant. For you to hear their stories out is 12.____

    A. *poor practice,* you should spend your time on more important matters
    B. *good practice,* this will increase your popularity with your subordinates
    C. *poor practice,* subordinates should learn to come to you only with major grievances
    D. *good practice,* it may prevent minor complaints from developing into major grievances

13. Assume that an agency has an established procedure for handling employee grievances. An employee in this agency, comes to his immediate supervisor with a grievance. The supervisor investigates the matter and makes a decision. 13.____
However, the employee is not satisfied with the decision made by the supervisor. The BEST action for the supervisor to take is to

    A. tell the employee he will review the matter further
    B. remind the employee that he is the supervisor and the employee must act in accordance with his decision
    C. explain to the employee how he can carry his complaint forward to the next step in the grievance procedure
    D. tell the employee he will consult with his own superiors on the matter

14. Subordinate employees and senior employees often must make quick decisions while in the field. The supervisor can BEST help subordinates meet such situations by 14.____

    A. training them in the appropriate action to take for every problem that may come up
    B. limiting the areas in which they are permitted to make decisions
    C. making certain they understand clearly the basic policies of the bureau and the department
    D. delegating authority to make such decisions to only a few subordinates on each level

15. Studies have shown that the CHIEF cause of failure to achieve success as a supervisor is 15.____

    A. an unwillingness to delegate authority to subordinates
    B. the establishment of high performance standards for subordinates
    C. the use of discipline that is too strict
    D. showing too much leniency to poor workers

16. When a supervisor delegates to a subordinate certain work that he normally does himself, it is MOST important that he give the subordinate 16.____

    A. responsibility for also setting the standards for the work to be done
    B. sufficient authority to be able to carry out the assignment
    C. written, step-by-step instructions for doing the work
    D. an explanation of one part of the task at a time

17. It is particularly important that disciplinary actions be equitable as between individuals. This statement *implies* that 17.____

A. punishment applied in disciplinary actions should be lenient
B. proposed disciplinary actions should be reviewed by higher authority
C. subordinates should have an opportunity to present their stories before penalties are applied
D. penalties for violations of the rules should be standardized and consistently applied

18. You discover that from time to time a number of false rumors circulate among your subordinates. 18.____
Of the following, the BEST way for you to handle this situation is to

A. ignore the rumors since rumors circulate in every office and can never be eliminated
B. attempt to find those responsible for the rumors and reprimand them
C. make sure that your employees are informed as soon as possible about all matters that affect them
D. inform your superior about the rumors and let him deal with the matter

19. Supervisors who allow the "halo effect" to influence their evaluations of subordinates are *most likely* to 19.____

A. give more lenient ratings to older employees who have longer service
B. let one highly favorable or unfavorable trait unduly affect their judgment of an employee
C. evaluate all employees on one trait before considering a second
D. give high evaluations in order to avoid antagonizing their subordinates

20. For a supervisor to keep records of reprimands to subordinates about infractions of the rules is 20.____

A. *good practice,* because these records are valuable to support disciplinary actions recommended or taken
B. *poor practice,* because such records are evidence of the supervisor's inability to maintain discipline
C. *good practice,* because such records indicate that the supervisor is doing a good job
D. *poor practice,* because the best way to correct subordinates is to give them more training

21. When a new departmental policy has been established, it would be MOST advisable for you, as a supervisor, to 21.____

A. distribute a memo which states the new policy and instruct your subordinates to read it
B. explain specifically to your subordinates how the policy is going to affect them
C. make sure your subordinates understand that you are not responsible for setting the policy
D. tell your subordinates whether you agree or disagree with the policy

22. As a supervisor, you receive several complaints about the rude conduct of a subordinate. The FIRST action you should take is to 22.____

    A. request his transfer to another office
    B. prepare a charge sheet for disciplinary action
    C. assign a senior employee to work with him for a week
    D. interview the employee to determine possible reason, and warn that correction is necessary

23. A supervisor is *most likely* to get subordinates to work cooperatively toward accomplishing bureau goals if he 23.____

    A. creates an atmosphere that contributes to their feeling of security
    B. backs up subordinates even when they occasionally disobey regulations
    C. shows interest in subordinates by helping them solve their personal problems
    D. uses an authoritarian or "bossy" approach to supervision

24. A supervisor is holding a staff meeting with his senior employees to try to find an acceptable solution to a problem that has come up. 24.____
    Of the following, the CHIEF role of the supervisor at this meeting should be to

    A. see that every member of the group contributes at least one suggestions
    B. act as chairman of the meeting, but take no other active part to avoid influencing the senior employees
    C. keep the participants from wandering off into discussions of irrelevant matters
    D. make certain the participants hear his views on the matter at the beginning of the meeting

25. An employee shows you a certificate that he has just received for completing two years of study in conversational Spanish. As his supervisor, it would be BEST for you to 25.____

    A. put a note about this accomplishment in his personnel folder
    B. assign him to areas in which people of Spanish origin live
    C. congratulate him on this accomplishment, but tell him frankly that you doubt this is likely to have any direct bearing on his work
    D. encourage him to continue his studies and become thoroughly fluent in speaking the language

KEY (CORRECT ANSWERS)

1. B
2. D
3. A
4. D
5. B
6. A
7. D
8. C
9. B
10. D
11. B
12. D
13. C
14. C
15. A
16. B
17. D
18. C
19. B
20. A
21. B
22. D
23. A
24. C
25. A

TEST 2

DIRECTIONS: Each question or incomplete statement is followed by several suggested answers or completions. Select the one that BEST answers the question or completes the statement. *PRINT THE LETTER OF THE CORRECT ANSWER IN THE SPACE AT THE RIGHT.*

1. Of the following, the factor affecting employee morale which the immediate supervisor is LEAST able to control is 1.____

   A. handling of grievances
   B. fair and impartial treatment of subordinates
   C. general presonnel rules and regulations
   D. accident prevention

2. When one of your workers does outstanding work, you should 2.____

   A. explain to your other employees that you expect the same kind of work of them
   B. praise him for his work so that he will know it is appreciated
   C. say nothing, because other employees may think you are showing favoritism
   D. show him how his work can be improved still more so that he will not sit back

3. For you as a supervisor to consider a suggestion from a probationary worker for improving a procedure would be 3.____

   A. *poor practice,* because this employee is too new on the job to know much about it
   B. *good practice,* because you may be able to share credit for the suggestion
   C. *poor practice,* because it may hurt the morale of the older employees
   D. *good practice,* because the suggestion may be worthwhile

4. If you find you must criticize the work of one of your workers, it would be BEST for you to 4.____

   A. mention the good points in his work as well as the faults
   B. caution him that he will receive an unsatisfactory performance report unless his work improves
   C. compare his work to that of the other agents you supervise
   D. apologize for making the criticism

5. As a senior employee which one of the following matters would it be BEST for you to talk over with your supervisor before you take final action? 5.____

   A. One of the workers you supervise continues to disregard your instructions repeatedly in spite of repeated warnings
   B. One of your workers tells you he wants to discuss a personal problem
   C. A probationary employee tells you he does not understand a procedure
   D. One of your workers tells you he disagrees with the way you rate his work

6. If one of your subordinates asks you a question about a department rule and you do not know the answer, you should tell him that 6.____

   A. he should try to get the information himself
   B. you do not have the answer, but you will get it for him as soon as you can
   C. he should ask you the question again a week from now
   D. he should put the question in writing

7. If, as a supervisor, you realize that you have been unfair in criticizing one of your subordinates, the BEST action for you to take is to 7.____

   A. say nothing, but overlook some error made by this employee in the future
   B. be frank and tell the employee that you are sorry for the mistake you made
   C. let the employee know in some indirect way without admitting your mistake, that you realize he was not at fault
   D. say nothing, but be more careful about criticizing subordinates in the future

8. Of the following, the MOST important reason for a supervisor to write an accident report as soon as possible after an accident has happened is to 8.____

   A. make sure that important facts about the accident are not forgotten
   B. avoid delay in getting compensation for the injured person
   C. get adequate medical treatment for the injured person
   D. keep department accident statistics up to date

9. In any matter which may require disciplinary action, the FIRST responsibility of the supervisor is to 9.____

   A. decide what penalty should be applied for the offense
   B. refer the matter to a higher authority for complete investigation
   C. place the interests of the department above those of the employee
   D. investigate the matter fully to get all the facts

10. Suppose you find it necessary to criticize one of the subordinates you supervise. You should 10.____

   A. send an official letter to his home
   B. speak to him about the matter privately
   C. speak to him at a staff meeting
   D. ask another worker who is friendly with him to talk to him about the matter

11. Some of your subordinates have been coming to you with complaints you feel are unimportant. For you to hear their stories out is 11.____

   A. *poor practice,* you should spend your time on more important matters
   B. *good practice,* this will increase your popularity with your subordinates
   C. *poor practice,* subordinates should learn to come to you only with major grievances
   D. *good practice,* it may prevent minor complaints from developing into major grievances

12. Suppose that NOT ONE of a group of employees has turned in an idea to the employees' suggestion system during the past year. The *most probable* reason for this situation is that the 12.____

   A. supervisor of these employees is not doing enough to encourage them to take part in this program
   B. employees in this group are not able to develop any good ideas
   C. money awards given for suggestions used are not high enough to make employees interested
   D. methods and procedures of operation do not need improvement

13. For you as a supervisor to give each of your subordinates *exactly* the same type of supervision is 13.____

A. *advisable,* because doing this insures fair and impartial treatment of each individual
B. *not advisable,* because each person is different and there is no one supervisory procedure for dealing with individuals that applies in every case
C. *advisable,* because once a supervisor learns how to deal with a subordinate who brings a problem to him, he can handle another subordinate with this problem in the same way
D. *not advisable,* because individuals like to think that they are receiving better treatment than others

14. In evaluating personnel, a supervisor should keep in mind that the MOST important objective of performance evaluations is to 14.____

A. encourage employees to compete for higher performance ratings
B. give recognition to employees who perform well
C. help employees improve their work
D. discipline employees who perform poorly

15. A subordinate tells you that he is having trouble concentrating on his work due to a personal problem at home. Of the following, it would be BEST for you to 15.____

A. refer him to a community service agency
B. listen quietly to the story because he may just need a sympathetic ear
C. tell him that you cannot help him because the problem is not job-related
D. ask him some questions about the nature of the problem and tell him how you would handle it

16. To do a good job of performance evaluation, it is BEST for a supervisor to 16.____

A. measure the employee's performance against standard performance requirements
B. compare the employee's performance to that of another employee doing similar work
C. leave out any consideration of the employee's personal traits
D. give greatest weight to instances of unusually good or unusually poor performance

17. It is particularly important that disciplinary actions be equitable as between individuals. This statement *implies* that 17.____

A. punishment applied in disciplinary actions should be lenient
B. proposed disciplinary actions should be reviewed by higher authority
C. subordinates should have an opportunity to present their stories before penalties are applied
D. penalties for violations of the rules should be standardized and consistently applied

18. Assume that an agency has an established procedure for handling employee grievances. An employee in this agency comes to his immediate supervisor with a grievance. The supervisor investigates the matter and makes a decision. However, the employee is not satisfied with the decision made by the supervisor.
The BEST action for the supervisor to take is to 18.____

A. tell the employee he will review the matter further
B. remind the employee that he is the supervisor and the employee must act in accordance with his decision
C. explain to the employee how he can carry his complaint forward to the next step in the grievance procedure
D. tell the employee he will consult with his own superiors on the matter

19. Of the following, the CHIEF purpose of a probationary period for a new employee is to allow time for 19.____

A. finding out whether the selection processes are satisfactory
B. determining the fitness of the employee to continue in the job
C. the employee to decide whether he wants a permanent appointment
D. the employee to make adjustments in his home circumstances made necessary by the job

20. Of the following, the subject that would be MOST important to include in a "break-in" program for new employees is 20.____

A. explanation of rules, regulations and policies of the agency
B. Instruction in the agency's history and programs
C. explanation of the importance of the new employees' own particular job
D. explanation of the duties and responsibilities of the employee

21. Suppose a new employee under your supervision seems slow to learn and is making mistakes in performing his duties. Your FIRST action should be to 21.____

A. pass this information on to the bureau director
B. reprimand the worker so he will not repeat these mistakes
C. find out whether this worker understands your instructions
D. note these facts for future reference when writing up the monthly performance evaluation

22. In training new employees to do a certain job it would be LEAST desirable for you to 22.____

A. demonstrate how the job is done, step by step
B. encourage the workers to ask questions if they aren't clear about any point
C. tell them about the various mistakes other agents have made in doing this job
D. have the workers do the job, explaining to you what they are doing and why

23. One of the workers under your supervision is resentful when you ask her to remove her jangling bracelets before she starts her tour of duty.
Of the following, the BEST explanation you can give her for the rule against wearing such jewelry while on duty is that 23.____

A. the jewelry may create a safety hazard
B. employees must give up certain personal liberties if they want to keep their jobs
C. workers cannot perform their duties as efficiently if they wear distracting jewelry
D. citizens may receive an unfavorable impression of the department

24. Of the following, the LEAST important reason for having a department handbook and a bureau standard operating procedure is to 24.____

    A. help in training new employees
    B. provide a source of reference for department and bureau rules and procedures
    C. prevent errors in work by providing clear guidelines
    D. make the supervisor's job easy

25. On inspecting your squad prior to their tour of duty, you note an employee improperly and unacceptably dressed. 25.____
    The FIRST action you should take is to

    A. call the employee aside and insist on immediate correction if possible
    B. notify the district commander right away
    C. have the employee submit a memorandum explaining the reason for the improper uniform
    D. permit the employee to proceed on duty but warn him not to let this happen again

---

KEY (CORRECT ANSWERS)

1.	C	11.	D
2.	B	12.	A
3.	D	13.	B
4.	A	14.	C
5.	A	15.	B
6.	B	16.	A
7.	B	17.	D
8.	A	18.	C
9.	D	19.	B
10.	B	20.	D

21. C
22. C
23. D
24. D
25. A

EXAMINATION SECTION
TEST 1

DIRECTIONS: Each question or incomplete statement is followed by several suggested answers or completions. Select the one that BEST answers the question or completes the statement. *PRINT THE LETTER OF THE CORRECT ANSWER IN THE SPACE AT THE RIGHT.*

1. The administrator who allows his staff to suggest ways to do their work will usually find that 1.____

    A. this practice contributes to high productivity
    B. the administrator's ideas produce greater output
    C. clerical employees suggest inefficient work methods
    D. subordinate employees resent performing a management function

2. In considering how to distribute among employees the various tasks which must be accomplished, an administrator should bear in mind that MOST people 2.____

    A. are working mainly for money, so the particular task they do is usually unimportant
    B. would rather work with a congenial group, but since this lowers output, it is better to have people work alone
    C. want recognition as outstanding workers, but since only one can be best, it is better policy to stress equality
    D. are concerned with being part of a group and also hope to be outstanding, and the administrator must consider both

3. A coordinator may be the supervisor of several employees. As such, he is their leader. The style of leadership which is MOST effective is a style in which 3.____

    A. the coordinator's behavior is tailored to the situation
    B. the coordinator lets his subordinates solve their own problems
    C. the coordinator consults with his subordinates about any work being done
    D. subordinates are told firmly what to do, how and when

4. As a coordinator you may be required to set up a records retention program. To set up such a program, the FIRST step you should take is to 4.____

    A. find out how long the records will be needed
    B. determine what types of records are maintained
    C. investigate storage facilities
    D. revise the filing system

5. You are responsible for supervising the work of several subordinates who deal directly with people seeking specific help from your department. One of your subordinates is faced with an angry citizen who has brought his troubles to the wrong department, but who refuses to believe this and is loudly demanding to *see the manager*. The subordinate asks you to step in and take over. Which of the following is probably the MOST effective way of handling this situation? 5.____

   A. Tell your subordinate that since this is obviously not a matter for your department, his request that you take over is inappropriate.
   B. Remind your subordinate firmly that it is his job to deal with the public, and that he must learn to handle people who are confused and angry.
   C. Do not stop what you are doing, but call out to the angry citizen that whatever your subordinate told him is correct.
   D. Step in and direct the angry citizen to a department which can help him.

6. A supervisor is one who is responsible for the actions of others working for him and at the same time is responsible to others above him in the organization chart. 6.____
The foregoing statement IMPLIES, in effect, that the supervisor

   A. has full authority for his actions
   B. can delegate his responsibilities to his assistants
   C. accepts direction from his own supervisor
   D. has higher status than a coordinator

7. In issuing requested supplies to employees of the office, there is a great deal of merit in limiting the quantity issued at any one time to about a two-week supply. 7.____
In MOST cases, this policy is

   A. *bad*, because employees should be allowed as large a quantity of supplies as they feel they need
   B. *bad*, because, if larger quantities were issued, employees would have to ask for supplies less often
   C. *good*, because the smaller the quantity issued, the more efficiently the office can be managed
   D. *good*, because a larger amount would encourage waste and a smaller amount would necessitate more trips to the stockroom

8. As a coordinator, assume that there is a rule in your office that all correspondence to other agencies must be signed personally by the hearing officer. 8.____
If the hearing officer is unexpectedly absent on a day when an important letter which has not yet been signed is scheduled to be mailed out, the MOST appropriate action for you to take is to

   A. seek advice from the superior of the hearing officer
   B. sign the letter with the name of the hearing officer and your own initials
   C. telephone the hearing officer at home
   D. wait until the next day

9. Unless otherwise directed, a car should be parked parallel to and within 12 inches of a curb or edge of a roadway, facing in the same direction as traffic on the car's side of the road. 9.____
Of the following, the MOST likely reason for this regulation is to

A. allow the car's passengers adequate room to open its doors
B. make sure that the road can be washed effectively by Sanitation Department equipment
C. prevent the car from blocking the smooth flow of traffic
D. allow another vehicle enough room to double-park

10. Parking meters are generally installed in shopping and commercial districts. 10.____
Of the following, the MOST likely reason for this practice is to

A. promote an equitable rotation of short-term parking opportunities
B. prevent trucks from stopping to unload and receive deliveries
C. discourage overnight parking of vehicles by local residents
D. maximize the revenue gained from these meters to offset the cost of purchasing and maintaining them

11. In the city, the large numbers of criminal cases have made it difficult for the court system to assure a defendant a speedy trial. 11.____
The MAIN result of this situation has been that

A. judges are imposing longer sentences to reduce the number of cases
B. defense attorneys and prosecutors often engage in plea bargaining
C. judges are being selected more rapidly by special *blue-ribbon* panels
D. juries are now given a limit of 48 hours within which they must deliver a verdict

12. The transit fare in the city may have to be raised to meet higher transit authority costs. 12.____
The one of the following which is MOST likely to be the PRINCIPAL cause of such higher costs is

A. equipment repair
B. equipment replacement
C. salary increases
D. conversion to air conditioning

Questions 13-16.

DIRECTIONS: Questions 13 through 16 are to be answered on the basis of the following table.

AVERAGE HOURLY CARRYING CAPACITIES OF
SINGLE-LANE TRANSPORT FACILITIES

MODE OF TRANSPORT	NO. OF PASSENGERS
Autos on surface streets	1,575
Autos on elevated highways	2,025
Buses on surface streets	9,000
Streetcars on surface streets	13,500
Streetcars in subways	20,000
Local subway trains	40,000
Express subway trains	60,000

13. For a group of elevated highways to approximately equal the carrying capacity of a two-lane local subway train facility, the TOTAL number of lanes required would be MOST NEARLY 13.____

A. 80 B. 60 C. 40 D. 20

14. Buses on surface streets using a single lane can carry approximately what percentage of the passengers that express subway trains in one lane can carry? 14.____

A. 20% B. 15% C. 10% D. 5%

15. The average number of passengers that can be carried by autos on surface streets in one day is MOST NEARLY 15.____

A. 1,575 B. 2,025 C. 37,800 D. 48,600

16. If one lane of a surface street were used for buses and another lane were used for streetcars, the number of passengers that could be carried by both lanes together in one hour would probably be MOST NEARLY 16.____

A. 9,000 B. 11,250 C. 13,500 D. 22,500

17. The one of the following which is MOST likely to result from a change from a centralized plan for records management to a decentralized plan is 17.____

A. a loss of time for personnel who use the records
B. greater specialization of record keeping personnel
C. authority and responsibility for the records management program being vested in one person within the organization
D. easier access to the records for personnel most concerned with such records

18. The *grapevine* is an informal means of communication in an organization. The attitude of a supervisor with respect to the grapevine should be to 18.____

A. ignore it since it deals mainly with rumors and sensational information
B. regard it as a serious danger which should be eliminated
C. accept it as a real line of communications which should be listened to
D. utilize it for most purposes instead of the official line of communication

19. The supervisor of an office that must deal with the public should realize that planning in this type of work situation 19.____

A. is useless because he does not know how many people will request service or what service they will request
B. must be done at a higher level but that he should be ready to implement the results of such planning
C. is useful primarily for those activities that are not concerned with public contact
D. is useful for all the activities of the office, including those that relate to public contact

20. Which of the following factors is MOST important in planning the location of work stations and other aspects of office layout? 20.____

A. Preferences of the office employees
B. Nature and flow of work in the office
C. Volume of work in the office
D. Seniority of employees in the office

KEY (CORRECT ANSWERS)

1. A	11. B
2. D	12. C
3. A	13. C
4. B	14. B
5. D	15. C
6. C	16. D
7. D	17. D
8. A	18. C
9. C	19. D
10. A	20. B

TEST 2

DIRECTIONS: Each question or incomplete statement is followed by several suggested answers or completions. Select the one that BEST answers the question or completes the statement. *PRINT THE LETTER OF THE CORRECT ANSWER IN THE SPACE AT THE RIGHT.*

1. It is usually MOST desirable for a work supervisor for a large group of clerical workers to have a work station which 1.____

   A. provides a view of the entire room
   B. is in another room away from all the clerical workers
   C. is isolated from all workers except for a secretary or assistant
   D. is located so that he can receive all visitors

Questions 2-5.

DIRECTIONS: Questions 2 through 5 must be answered on the basis of the following passage.

Analysis of current data reveals that motor vehicle transportation actually requires less space than was used for other types of transportation in the pre-automobile era, even including the substantial area taken by freeways. The reason is that when the fast moving through traffic is put on built-for-the-purpose arterial roads, then the amount of ordinary space needed for strictly local movement and for access to property drops sharply. Even the amount of land taken for urban expressways turns out to be surprisingly small in terms either of total urban acreage or of the volume of traffic they carry. No existing or contemplated urban expressway system requires as much as 3 percent of the land in the areas it serves, and this would be exceptionally high. The Los Angeles freeway system occupies only 2 percent of the available land; the same is true of the District of Columbia, where only 0.75 percent is pavement, with the remaining 1.25 percent as open space. California studies estimate that, in a typical California urban community, 1.6 to 2 percent of the area should be devoted to freeways, which will handle 50 to 60 percent of all traffic needs, and about ten times as much land to the ordinary roads and streets that carry the rest of the traffic. By comparison, when John A. Sutter laid out Sacramento in 1850, he provided 38 percent of the area for streets and sidewalks. The French architect, Pierre L' Enfant, proposed 59 percent of the area of the District of Columbia for roads and streets; urban renewal in Southwest Washington, incorporating a modern street network, reduced the acreage of space for pedestrian and vehicular traffic in the renewal area from 48.2 to 41.5 percent of the total. If we are to have a reasoned consideration of the impact of highway transportation on contemporary urban development, it would be well to understand these relationships.

2. The passage states that 2.____

   A. modern transportation uses less space than was used for transportation before the auto age
   B. expressways require more space than streets in terms of urban acreage
   C. typical urban communities were poorly designed in terms of relationship between space used for traffic and that used for other purposes
   D. the need for local and access roads would increase if the number of expressways were increased

3. According to the above passage, it was originally planned that the percent of the area to be used for roads and streets in the District of Columbia should be MOST NEARLY 3.____

   A. 40% B. 45% C. 50% D. 60%

4. The above passage states that the amount of space needed for local traffic 4.____

A. *increases* when arterial highways are constructed
B. *decreases* when arterial highways are constructed
C. *decreases* when there is more land available
D. *increases* when there is more land available

5. According to the above passage, studies estimate that, in a typical California urban community, the amount of land devoted to ordinary roads and streets as compared with that devoted to freeways should be MOST NEARLY ______ as much. 5.____

A. one-half
B. one-tenth
C. twice
D. ten times

Questions 6–8.

DIRECTIONS: Questions 6 through 8 must be answered on the basis of the following passage.

A glaring exception to the usual practice of the judicial trial as a means of conflict resolution is the utilization of administrative hearings. The growing tendency to create administrative bodies with rule-making and quasi-judicial powers has shattered many standard concepts. A comprehensive examination of the legal process cannot neglect these newer patterns.

In the administrative process, the legislative, executive, and judicial functions are mixed together, and many functions, such as investigating, advocating, negotiating, testifying, rule-making, and adjudicating, are carried out by the same agency. The reason for the breakdown of the separation-of-powers formula is not hard to find. It was felt by Congress, and state and municipal legislatures, that certain regulatory tasks could not be performed efficiently, rapidly, expertly, and with due concern for the public interest by the traditional branches of government. Accordingly, regulatory agencies were delegated powers to consider disputes from the earliest stage of investigation to the final stages of adjudication entirely within each agency itself, subject only to limited review in the regular courts.

6. The above passage states that the usual means for conflict resolution is through the use of 6.____

A. judicial trial
B. administrative hearing
C. legislation
D. regulatory agencies

7. The above passage *implies* that the use of administrative hearing in resolving conflict is a(n) ______ approach. 7.____

A. traditional
B. new
C. dangerous
D. experimental

8. The above passage states that the reason for the breakdown of the separation-of-powers formula in the administrative process is that 8.____

A. Congress believed that certain regulatory tasks could be better performed by separate agencies
B. legislative and executive functions are incompatible in the same agency
C. investigative and regulatory functions are not normally reviewed by the courts
D. state and municipal legislatures are more concerned with efficiency than with legality

9. An employee examining the summonses of individuals appearing for hearings noticed that the address on one summons was the same as that of an individual who had appeared earlier that day. He asked the second respondent if he knew the first respondent.
The MOST appropriate evaluation of the employee's behavior is that he should 9.____

   A. not have mentioned any other respondent to the second respondent
   B. not waste time inspecting summonses in such detail
   C. be commended for inspecting summonses so carefully
   D. be commended for his investigation of the respondents

10. An employee is assigned to maintain all types of frequently used reference material such as booklets and technical papers. He keeps these in a pile on a shelf in order of arrival. When new material arrives, he puts it on top of the pile.
Which of the following BEST evaluates the employee's handling of this reference material?
His system is most likely to result in______ filing and ______ retrieval. 10.____

   A. *fast; slow*
   B. *slow; slow*
   C. *fast; fast*
   D. *slow; fast*

11. An employee computes statistics relating to proceedings. The method he devised consists of organizing his source and summary documents in such a manner that at any time another employee can assume the work. This method takes a little more time than other possible methods.
Which of the following statements BEST evaluates the judgment of the employee in devising such a method?
The employee has used 11.____

   A. *good* judgment because it is important to provide for continuity
   B. *poor* judgment because he is not using the fastest method
   C. *good* judgment because if a job is done as fast as possible, it becomes tiring
   D. *poor* judgment because it is not an employee's responsibility to prepare for a replacement

12. Assume that it is your job to receive incoming telephone calls. Those calls which you cannot handle yourself have to be transferred to the appropriate office.
If you receive an outside call for an extension line which is busy, the one of the following which you should do FIRST is to 12.____

   A. interrupt the person speaking on the extension and tell him a call is waiting
   B. tell the caller the line is busy and let him know every thirty seconds whether or not it is free
   C. leave the caller on *hold* until the extension is free
   D. tell the caller the line is busy and ask him if he wishes to wait

13. On one occasion in a certain office, an elderly employee collapsed, apparently the victim of a heart attack. Chaos broke out in the office as several people tried to help him, and several others tried to get assistance. 13.____
Of the following, the MOST certain way of avoiding such chaos in the future is to

A. keep a copy of heart attack procedures on file so that it can be referred to by any member of the staff when an emergency occurs
B. provide each member of the staff with a first aid book which is to be kept in an accessible location
C. train all members of the staff in the proper procedure for handling such emergencies, assigning specific responsibilities
D. post, in several places around the office, a list of specific procedures to follow in each of several different emergencies

14. Your superior has subscribed to several publications directly related to your division's work, and he has asked you to see to it that the publications are circulated among the supervisory personnel in the division. There are eight supervisors involved. 14.____
The BEST method of insuring that all eight see these publications is to

A. place the publication in the division's general reference library as soon as it arrives
B. inform each supervisor whenever a publication arrives and remind all of them that they are responsible for reading it
C. prepare a standard slip that can be stapled to each publication, listing the eight supervisors and saying, *Please read, initial your name, and pass along*
D. send a memo to the eight supervisors saying that they may wish to purchase individual subscriptions in their own names if they are interested in seeing each issue

15. Assume that you have been asked to prepare a narrative summary of the monthly reports submitted by employees in your division. 15.____
In preparing your summary of this month's reports, the FIRST step to take is to

A. read through the reports, noting their general content and any unusual features
B. decide how many typewritten pages your summary should contain
C. make a written summary of each separate report, so that you will not have to go back to the original reports again
D. ask each employee which points he would prefer to see emphasized in your summary

16. Your superior has telephoned a number of key officials in your agency to ask whether they can meet at a certain time next month. He has found that they can all make it, and he has asked you to confirm the meeting. 16.____
Which of the following is the BEST way to confirm such a meeting?

A. Note the meeting on your superior's calendar
B. Post a notice of the meeting on the agency bulletin board
C. Call the officials on the day of the meeting to remind them of the meeting
D. Write a memo to each official involved repeating the time and place of the meeting

17. Of the following, the worker who is MOST likely to create a problem in maintaining safety is one who 17.____

A. disregards hazards
B. feels tired
C. resents authority
D. gets bored

18. Assume that a new regulation requires that certain kinds of private organizations file information forms with your department. You have been asked to write the short explanatory message that will be printed on the front cover of the pamphlet containing the forms and instructions. 18.____
Which of the following would be the MOST appropriate way of beginning this message?

A. Get the readers' attention by emphasizing immediately that there are legal penalties for organizations that fail to file before a certain date
B. Briefly state the nature of the enclosed forms and the types of organizations that must file
C. Say that your department is very sorry to have to put organizations to such an inconvenience
D. Quote the entire regulation adopted by the city, even if it is quite long and is expressed in complicated legal language

19. Suppose that you have been told to make up the vacation schedule for the 15 employees in a particular unit. In order for the unit to operate effectively, only a few employees can be on vacation at the same time. 19.____
Which of the following is the MOST advisable approach in making up the schedule?

A. Draw up a schedule assigning vacations in alphabetical order
B. Find out when the supervisors want to take their vacations, and randomly assign whatever periods are left to the non-supervisory personnel
C. Assign the most desirable times to employees of longest standing, and the least desirable times to the newest employees
D. Have all employees state their own preference, and then work out any conflicts in consultation with the people involved

20. Assume that you have been asked to prepare job descriptions for various positions in your department. 20.____
Which of the following are the BASIC points that should be covered in a job description?

A. General duties and responsibilities of the position, with examples of day-to-day tasks
B. Comments on the performances of present employees
C. Estimates of the number of openings that may be available in each category during the coming year
D. Instructions for carrying out the specific tasks assigned to your department

KEY (CORRECT ANSWERS)

1.	A	11.	A
2.	A	12.	D
3.	D	13.	C
4.	B	14.	C
5.	D	15.	A
6.	A	16.	D
7.	B	17.	A
8.	A	18.	B
9.	A	19.	D
10.	A	20.	A

EXAMINATION SECTION
TEST 1

DIRECTIONS: Each question or incomplete statement is followed by several suggested answers or completions. Select the one that BEST answers the question or completes the Statement. *PRINT THE LETTER OF THE CORRECT ANSWER IN THE SPACE AT THE RIGHT.*

1. Your superior has asked you to notify division employees of an important change in one of the operating procedures described in the division manual. Every employee presently has a copy of this manual. 1.____
Which of the following is *normally* the MOST practical way to get the employees to understand such a change?

   A. Notify each employee individually of the change and answer any questions he might have
   B. Send a written notice to key personnel, directing them to inform the people under them
   C. Call a general meeting, distribute a corrected page for the manual, and discuss the change
   D. Send a memo to employees describing the change in general terms and asking them to make the necessary corrections in their copies of the manual

2. A supervisor was directed by the head of his division to report figures for overtime wages. The supervisor asked a clerk under his supervision to give him the figures, and he passed the clerk's figures along to his superior without questioning them. It was then discovered that the clerk had carelessly supplied the wrong information. Who can PROPERLY be held responsible for the mistake, the supervisor or the payroll clerk? 2.____

   A. Only the supervisor, because he should have known that the clerk would be careless
   B. Only the clerk, because it should be unnecessary for supervisors to check the work of their subordinates except for work which is unusually complex or important
   C. Neither of them, because it is perfectly understandable that such mistakes will occur from time to time
   D. Both of them, because the person to whom a task is delegated is responsible to the supervisor who delegated the task, and the supervisor is responsible to his superior

3. As a supervisor, it is necessary for you to show a new employee how to enter information on standard forms that he will have to prepare. These forms have a number of blanks to be filled in, but the job is fairly simple once a person becomes familiar with it. 3.____
The BEST way to show the new employee how to do the job is to

   A. explain how to do it and have him fill out a few forms, helping him with any difficulties
   B. give him a completed form to use as a model, and tell him to do all the others exactly the same way
   C. put him on his own immediately, and assume that he will learn for himself through trial and error
   D. give him several dozen completed forms to read, and ask him to check back with you in a few hours when he feels ready to start work

4. An administrative position carries with it a certain amount of authority. Management theorists feel that the exercise of authority is ESSENTIAL in carrying out the goals of an organization because 4.____

   A. administrators enjoy having the power to order people around, and they would not be willing to give it up
   B. administrators must work through others to accomplish objectives, so they must have the right to direct others to act in certain ways
   C. most employees are not able to carry out tasks on their own initiative, and they need a stern supervisor to make sure that the work gets done
   D. once authority is recognized, it can be carefully limited so that no administrator makes unreasonable demands or sets himself up as a petty tyrant

5. Assume that the work in your department involves the use of many technical terms. In such a situation, when you are answering inquiries from the general public, it would *usually* be BEST to 5.____

   A. use simple language and avoid the technical terms
   B. use the technical terms whenever possible
   C. use technical terms freely, but explain each term in parentheses
   D. apologize if you are forced to use a technical term

6. You are answering a letter that was written on the letterhead of the ABC Company and signed by James H. Block, Treasurer. What is usually considered to be the CORRECT salutation to use in your reply? 6.____
   Dear

   A. ABC Company:
   B. Sirs:
   C. Mr. Block:
   D. Mr. Treasurer:

7. Assume that one of your duties is to handle routine letters of inquiry from the public. The one of the following which is *usually* considered to be MOST desirable in replying to such a letter is a 7.____

   A. detailed answer handwritten on the original letter of inquiry
   B. phone call, since you can cover details more easily over the phone than in a letter
   C. short letter giving the specific information requested
   D. long letter discussing all possible aspects of the questions raised

8. The CHIEF reason for dividing a letter into paragraphs is to 8.____

   A. make the message clear to the reader by starting a new paragraph for each new topic
   B. make a short letter occupy as much of the page as possible
   C. keep the reader's attention by providing a pause from time to time
   D. make the letter look neat and business like

9. Your superior has asked you to send a letter via fax from your agency to a government agency in another city. He has written the letter and provided you with all contact information. 9.____
Which of the following does not need to be included on the fax cover sheet to be sent along with your superior's letter?

A. Today's date
B. A final sentence such as, *We would appreciate hearing from your agency in reply as soon as is convenient for you*
C. Name of the contact person or department at the other agency
D. Name of sender

10. Suppose that a usually competent employee whom you supervise has suddenly begun having difficulty completing his assignments. You ask the employee to speak to you privately about this situation and he agrees that he would appreciate this opportunity because of a problem he is having. 10.____
Of the following, which one would be the BEST technique for you to use in speaking with him?

A. Criticize the employee's performance as soon as he mentions his difficulty in completing his assignments
B. Listen patiently to what the employee has to say before making any comments on your own
C. Refuse to discuss any personal factors which the employee mentions when he tries to explain his recent work difficulty
D. Allow the employee to argue with you but plan your attack and defense carefully

11. Suppose that you receive a telephone call from someone identifying himself as an employee in another department who asks to be given information which your own department regards as confidential. 11.____
Which of the following is the BEST way of handling such a request?

A. Give the information requested, since your caller has official standing
B. Grant the request, provided the caller gives you a signed receipt
C. Refuse the request, because you have no way of knowing whether the caller is really who he claims to be
D. Explain that the information is confidential and inform the caller of the channels he must go through to have the information released to him

12. The MAIN purpose of transferring materials from active to inactive files is to 12.____

A. keep current reference files from growing to a size where they become inefficient and unmanageable
B. distinguish between important business and less important matters
C. provide a means of storing letters that need not be answered
D. make sure that there is some way of retrieving information from previous years

13. The one of the following for which a cross-index is MOST likely to be needed is a 13.____

A. file of reference material arranged by subject
B. file of individual personnel records arranged alphabetically
C. card file containing addresses and phone numbers for various organizations
D. supervisor's tickler file

14. The CHIEF advantage of a rotary file is that 14.____

A. it holds much more material than a standard file cabinet
B. it provides a temporary location for material that is due to be placed in the permanent files
C. items can be easily located and scanned without being removed from the file
D. less time is required for placing an item on a rotary file than for placing it in a standard upright file

15. Centralization of office activities has become an important technique for achieving greater efficiency in clerical work. 15.____
Which of the following is NOT a result that could *normally* be gained by centralization of a clerical activity?

A. More even distribution of work loads among employees performing the same kind of clerical tasks
B. Increased opportunities for clerical workers to learn new skills and become better qualified for promotion to administrative positions
C. Cost savings on office equipment whose use can now be shared by several employees
D. Establishment of uniform standards and procedures for various clerical activities

16. Assume that certain work processed in your office is then sent to another office for further processing. One of the employees in your office tells you that the supervisor in the other office has been complaining about your office's method of handling the work. 16.____
Of the following, the MOST appropriate action for you to take is to

A. get all the details from the employee and then speak to the other supervisor
B. ignore the situation and continue to do the best you can
C. remind the supervisor that it is not his function to evaluate your work
D. refrain from reporting the matter to your superior

17. It is the practice in your department to make objective evaluations of the performance of different units. This requires looking at the results achieved by a particular unit during a specified period of time–for instance, the number of applications processed, the number of inquiries answered, the number of inspections made, and so forth. 17.____
Of the following, the BEST method of evaluating the performance of each unit is to compare its results With the

A. results achieved by all units of the same size that are performing other kinds of work
B. goals that the unit was reasonably expected to meet during the specified period
C. performance of the same unit during a similar period of time four or five years earlier
D. amount of money spent to achieve these results

18. It is possible that you may be asked to submit a brief written evaluation of the work of several employees under your supervision. Such an evaluation should *normally* give LEAST emphasis to an employee's 18.____

A. attendance record, including tardiness and absence
B. ability to grasp new assignments and carry them out effectively
C. educational background and previous employment experience
D. ability to get along with co-workers

19. You have been asked to help draw up a plan for a new operation to be carried out by your department. The INITIAL step in planning should be 19.____
    A. finding out how much money is available in the budget for the operation
    B. determining the objective or objectives of the operation
    C. gathering information on similar operations elsewhere
    D. determining the most reasonable way of structuring the operation

20. Studies show that office employees place high importance on the social and human aspects of the organization. What office employees like best about their jobs is the kind of people with whom they work. So strive hard to group people who are most likely to get along well together. 20.____
    Based on this statement, it is MOST reasonable to assume that office workers are most pleased to work in a group which
    A. is congenial
    B. has high productivity
    C. allows individual creativity
    D. is unlike other groups

KEY (CORRECT ANSWERS)

1.	C	11.	D
2.	D	12.	A
3.	A	13.	A
4.	B	14.	C
5.	A	15.	B
6.	C	16.	A
7.	C	17.	B
8.	A	18.	C
9.	B	19.	B
10.	B	20.	A

TEST 2

DIRECTIONS: Each question or incomplete statement is followed by several suggested answers or completions. Select the one that BEST answers the question or completes the statement. *PRINT THE LETTER OF THE CORRECT ANSWER IN THE SPACE AT THE RIGHT.*

1. A certain coordinator does not compliment members of his staff when they come up with good ideas. He feels that coming up with good ideas is part of the job and does not merit special attention. 1.____
This coordinator's practice is

   A. *poor,* because recognition for good ideas is a good motivator
   B. *poor,* because the staff will suspect that the coordinator has no good ideas of his own
   C. *good,* because it is reasonable to assume that employees will tell their supervisor of ways to improve office practice
   D. *good,* because the other members of the staff are not made to seem inferior by comparison

2. An employee under your supervision complains about a decision you have made in assigning work in the office. You consider the matter to be unimportant, but it seems to be very important to him. He is excited and very angry. 2.____
Of the following, the MOST appropriate action for you to take FIRST is to

   A. listen to the details of his complaint
   B. refer him to your superior
   C. tell him to *cool off* before discussing the matter
   D. tell him to settle it with the other employees

3. An experienced employee complains to his unit supervisor that the latter's continual, very close supervision of his work is unnecessary and annoying. The unit supervisor has been recently appointed. 3.____
Of the following, it would generally be BEST for the unit supervisor to

   A. agree to discontinue all supervision if the employee will agree, if he has any problems, to consult the supervisor
   B. assure the employee that close supervision is necessary but should not be taken personally
   C. consider with the employee what aspects of the supervision could be reduced
   D. explain that he is supervising closely only until he learns what the job is all about

4. A coordinator had a clerk assigned to help him review records. One day the coordinator asked the clerk to continue checking the records, and the clerk said, *No, I'm not doing any more of that today.* 4.____
In this instance, the coordinator should IMMEDIATELY

   A. ask the clerk why he will not check the records
   B. ask another clerk to do the job
   C. tell the clerk he must do it or be transferred
   D. contact his own supervisor

5. Assume that you have been assigned to supervise other employees. You find that one of your subordinates makes many mistakes whenever he prepares a particular report. Of the following, the MOST desirable course of action for you to follow FIRST in such a situation is to 5.____

   A. retrain the subordinate in the preparation of the report
   B. transfer the subordinate to another unit
   C. tell the subordinate to improve or resign
   D. give the employee different duties

6. Some employees of a department have sent an anonymous letter containing many complaints to the department head. Of the following, what is this MOST likely to show about the department? 6.____

   A. It is probably a good place to work.
   B. Communications are probably poor.
   C. The complaints are probably unjustified.
   D. These employees are probably untrustworthy.

7. Of the following, the BEST reason for rotating employee work assignments is that such rotation 7.____

   A. challenges the ingenuity of supervisors in making assignments
   B. gives each employee a chance at both desirable and undesirable assignments
   C. creates specialists among all employees
   D. increases the competitive spirit among employees

8. A citizen was angry about a parking ticket which he had received, and he insisted on talking to a coordinator about a hearing. The coordinator spoke to him and explained the rules and procedures relating to the disposition of summonses for parking violations. The citizen remained angry and dissatisfied. The coordinator then appealed to the citizen's civic responsibility and asked him if he wished to be an obstructionist. This last action was incorrect. 8.____
   How should the coordinator have handled this situation?

   A. Summoned a supervisor immediately and not talked with the angry citizen
   B. Been more sympathetic and shown some agreement with the citizen's complaint
   C. Limited himself to explaining the rules and regulations
   D. Shown some anger himself in order to reduce the citizen's anger

9. A coordinator has had several problems with a clerk who assists him. He calls the clerk in for a discussion of the matters. 9.____
   Which of the following should comprise the MAJOR part of the discussion?

   A. All the things the clerk has done wrong
   B. The most recent things the clerk has done wrong
   C. The things the clerk has done well in addition to the things he has done wrong
   D. The clerk's previous experience and personal problems

Questions 10–14.

DIRECTIONS: Questions 10 through 14 are to be answered SOLELY on the basis of the following passage.

The laws with which criminal courts are concerned contain threats of punishment for infraction of specified rules. Consequently, the courts are organized primarily for implementation of the punitive societal reaction of crime. While the informal organization of most courts allows the judge to use discretion as to which guilty persons actually are to be punished, the threat of punishment for all guilty persons always is present. Also, in recent years a number of formal provisions for the use of non-punitive and treatment methods by the criminal courts have been made, but the threat of punishment remains, even for the recipients of the treatment and non-punitive measures. For example, it has become possible for courts to grant probation, which can be non-punitive, to some offenders, but the probationer is constantly under the threat of punishment, for, if he does not maintain the conditions of his probation, he may be imprisoned. As the treatment reaction to crime becomes more popular, the criminal courts may have as their sole function the determination of the guilt or innocence of the accused persons, leaving the problem of correcting criminals entirely to outsiders. Under such conditions, the organization of the court system, the duties and activities of court personnel, and the nature of the trial all would be decidedly different.

10. Which one of the following is the BEST description of the subject matter of the above passage? The 10.____

A. value of non-punitive measures for criminals
B. effect of punishment on guilty individuals
C. punitive functions of the criminal courts
D. success of probation as a deterrent of crime

11. It may be INFERRED from the above passage that the present traditional organization of the criminal court system is a result of 11.____

A. the nature of the laws with which these courts are concerned
B. a shift from non-punitive to punitive measures for correctional purposes
C. an informal arrangement between court personnel and the government
D. a formal decision made by court personnel to increase efficiency

12. All persons guilty of breaking certain specified rules, according to the above passage, are subject to the threat of 12.____

A. treatment
B. punishment
C. probation
D. retrial

13. According to the above passage, the decision whether or not to punish a guilty person is a function USUALLY performed by 13.____

A. the jury
B. the criminal code
C. the judge
D. corrections personnel

14. According to the above passage, which one of the following is a possible effect of an increase in the *treatment reactions to crime*? 14.____

    A. A decrease in the number of court personnel
    B. An increase in the number of criminal trials
    C. Less reliance on probation as a non-punitive treatment measure
    D. A decrease in the functions of the court following determination of guilt

15. Which of the following actions would usually be MOST appropriate for a coordinator to take after receiving an instruction sheet from his supervisor explaining a new procedure which is to be followed? 15.____

    A. Put the instruction sheet aside temporarily until he determines what is wrong with the old procedure
    B. Call his supervisor and ask if the procedure is one he must implement immediately
    C. Write a memorandum to the supervisor asking for more details
    D. Try the new procedure and advise the supervisor of any problem or possible improvements

16. Assume that you are in charge of an office handling a large volume of various types of clerical work. 16.____
    The one of the following that must be done FIRST to promote even distribution and proper flow of work is to determine

    A. when additional work will come to the office
    B. the capabilities of the staff
    C. the type of tasks to be done and their priorities
    D. the time required for each task

17. In a miscellaneous correspondence folder in a file drawer, it is usually MOST helpful if letters are arranged according to 17.____

    A. date with the most recent date on the bottom
    B. date with the most recent date on the top
    C. subject with the subjects alphabetically arranged
    D. name with the names arranged geographically

18. Of the following, which one is considered the PRIMARY advantage of using a committee to resolve a problem in an organization? 18.____

    A. No one person will be held accountable for the decision since a group of people was involved
    B. People with different backgrounds give attention to the problem
    C. The decision will take considerable time so there is unlikely to be a decision that will later be regretted
    D. One person cannot dominate the decision-making process

19. Assume that as a coordinator you have been asked to redesign a form used in your office. 19.____
    Of the following, your MOST important consideration should be the

    A. sequence of items on the form
    B. number of items to be included on the form
    C. number of copies that are required
    D. purpose of the form

20. Employees in a certain office come to their supervisor with all their complaints about the office and the work. Almost every employee has had at least one minor complaint at some time. 20.____
The situation with respect to complaints in this office may BEST be described as probably

A. *good;* employees who complain care about their jobs and work hard
B. *good;* grievances brought out into the open can be corrected
C. *bad;* only serious complaints should be discussed
D. *bad;* it indicates the staff does not have confidence in the administration

KEY (CORRECT ANSWERS)

1. A
2. A
3. C
4. A
5. A
6. B
7. B
8. C
9. C
10. C
11. A
12. B
13. C
14. D
15. D
16. C
17. B
18. B
19. D
20. B

PREPARING WRITTEN MATERIAL

PARAGRAPH REARRANGEMENT COMMENTARY

The sentences that follow are in scrambled order. You are to rearrange them in proper order and indicate the letter choice containing the correct answer at the space at the right.

Each group of sentences in this section is actually a paragraph presented in scrambled order. Each sentence in the group has a place in that paragraph; no sentence is to be left out. You are to read each group of sentences and decide upon the best order in which to put the sentences so as to form a well-organized paragraph.

The questions in this section measure the ability to solve a problem when all the facts relevant to its solution are not given.

More specifically, certain positions of responsibility and authority require the employee to discover connection between events sometimes, apparently, unrelated. In order to do this, the employee will find it necessary to correctly infer that unspecified events have probably occurred or are likely to occur. This ability becomes especially important when action must be taken on incomplete information.

Accordingly, these questions require competitors to choose among several suggested alternatives, each of which presents a different sequential arrangement of the events. Competitors must choose the MOST logical of the suggested sequences.

In order to do so, they may be required to draw on general knowledge to infer missing concepts or events that are essential to sequencing the given events. Competitors should be careful to infer only what is essential to the sequence. The plausibility of the wrong alternatives will always require the inclusion of unlikely events or of additional chains of events which are NOT essential to sequencing the given events.

It's very important to remember that you are looking for the best of the four possible choices, and that the best choice of all may not even be one of the answers you're given to choose from.

There is no one right way to solve these problems. Many people have found it helpful to first write out the order of the sentences, as they would have arranged them, on their scrap paper before looking at the possible answers. If their optimum answer is there, this can save them some time. If it isn't, this method can still give insight into solving the problem. Others find it most helpful to just go through each of the possible choices, contrasting each as they go along. You should use whatever method feels comfortable and works for you.

While most of these types of questions are not that difficult, we've added a higher percentage of the difficult type, just to give you more practice. Usually there are only one or two questions on this section that contain such subtle distinctions that you're unable to answer confidently. And you then may find yourself stuck deciding between two possible choices, neither of which you're sure about.

EXAMINATION SECTION

TEST 1

DIRECTIONS: The sentences that follow are in scrambled order. You are to rearrange them in proper order and indicate the letter choice containing the correct answer. *PRINT THE LETTER OF THE CORRECT ANSWER IN THE SPACE AT THE RIGHT.*

1. Below are four statements labeled W, X, Y and Z. 1.____
   W. He was a strict and fanatic drillmaster.
   X. The word is always used in a derogatory sense and generally shows resentment and anger on the part of the user.
   Y. It is from the name of this Frenchman that we derive our English word, martinet.
   Z. Jean Martinet was the Inspector-General of Infantry during the reign of King Louis XIV.

   The PROPER order in which these sentences should be placed in a paragraph is:
   A. X, Z, W, Y B. X, Z, Y, W C. Z, W, Y, X D. Z, Y, W, X

2. In the following paragraph, the sentences, which are numbered, have been jumbled. 2.____
   I. Since then it has undergone changes.
   II. It was incorporated in 1955 under the laws of the State of New York.
   III. Its primary purposes, a cleaner city, has, however, remained the same.
   IV. The Citizens Committee works in cooperation with the Mayor's Inter-departmental Committee for a Clean City. 3.____

   The order in which these sentences should be arranged to form a well-organized paragraph is:
   A. II, IV, I, III B. III, IV, I, II C. IV, II, I, III D. IV, III, II, I

Questions 3-5.

DIRECTIONS: The sentences listed below are part of a meaningful paragraph but they are not given in their proper order. You are to decide what would be the BEST order in which to put the sentences so as to form a well-organized paragraph. Each sentence has a place in the paragraph; there are no extra sentences. You are then to answer Questions 3 through 5 inclusive on the basis of your rearrangements of these scrambled sentences into a properly organized paragraph.

In 1887 some insurance companies organized an Inspection Department to advise their clients on all phases of fire prevention and protection. Probably this has been due to the smaller annual fire losses in Great Britain than in the United States. It tests various fire prevention devices and appliances and determines manufacturing hazards and their safeguards. Fire research began earlier in the United States and is more advanced than in Great Britain. Later they established a laboratory specializing in electrical, mechanical, hydraulic, and chemical fields.

3. When the five sentences are arranged in proper order, the paragraph starts with the sentence which begins 3.____
   A. "In 1887..." B. "Probably this..." C. "It tests..."
   D. "Fire research..." E. "Later they..."

4. In the last sentence listed above, "they" refers to 4.____
   A. the insurance companies
   B. the United States and Great Britain
   C. the Inspection Department
   D. clients
   E. technicians

5. When the above paragraph is properly arranged, it ends with the words 5.____
   A. "...and protection."
   B. "...the United States."
   C. "...their safeguards."
   D. "...in Great Britain."
   E. "...chemical fields."

KEY (CORRECT ANSWERS)

1. C
2. C
3. D
4. A
5. C

TEST 2

DIRECTIONS: In each of the questions numbered I through V, several sentences are given. For each question, choose as your answer the group of number that represents the MOST logical order of these sentences if they were arranged in paragraph form. *PRINT THE LETTER OF THE CORRECT ANSWER IN THE SPACE AT THE RIGHT.*

1. I. It is established when one shows that the landlord has prevented the tenant's enjoyment of his interest in the property leased. 1.____
   II. Constructive eviction is the result of a breach of the covenant of quiet enjoyment implied in all leases.
   III. In some parts of the United States, it is not complete until the tenant vacates within a reasonable time.
   IV. Generally, the acts must be of such serious and permanent character as to deny the tenant the enjoyment of his possessing rights.
   V. In this event, upon abandonment of the premises, the tenant's liability for that ceases.

   The CORRECT answer is:
   A. II, I, IV, III, V B. V, II, III, I, IV
   C. IV, III, I, II, V D. I, III, V, IV, II

2. I. The powerlessness before private and public authorities that is the typical experience of the slum tenant is reminiscent of the situation of blue-collar workers all through the nineteenth century. 2.____
   II. Similarly, in recent years, this chapter of history has been reopened by anti-poverty groups which have attempted to organize slum tenants to enable them to bargain collectively with their landlords about the conditions of their tenancies.
   III. It is familiar history that many of the worker remedied their condition by joining together and presenting their demands collectively.
   IV. Like the workers, tenants are forced by the conditions of modern life into substantial dependence on these who possess great political aid and economic power.
   V. What's more, the very fact of dependence coupled with an absence of education and self-confidence makes them hesitant and unable to stand up for what they need from those in power.

   The CORRECT answer is:
   A. V, IV, I, II, III B. II, III, I, V, IV
   C. III, I, V, IV, II D. I, IV, V, III, II

3. I. A railroad, for example, when not acting as a common carrier may contract away responsibility for its own negligence. 3.____
   II. As to a landlord, however, no decision has been found relating to the legal effect of a clause shifting the statutory duty of repair to the tenant.
   III. The courts have not passed on the validity of clauses relieving the landlord of this duty and liability.
   IV. They have, however, upheld the validity of exculpatory clauses in other types of contracts.

V. Housing regulations impose a duty upon the landlord to maintain leased premises in safe condition.
VI. As another example, a bailee may limit his liability except for gross negligence, willful acts, or fraud.

The CORRECT answer is:

A. II, I, VI, IV, III, V
B. I, III, IV, V, VI, II
C. III, V, I, IV, II, VI
D. V, III, IV, I, VI, II

4. I. Since there are only samples in the building, retail or consumer sales are generally eschewed by mart occupants, and in some instances, rigid controls are maintained to limit entrance to the mart only to those persons engaged in retailing. 4.____
II. Since World War I, in many larger cities, there has developed a new type of property, called the mart building.
III. It can, therefore, be used by wholesalers and jobbers for the display of sample merchandise.
IV. This type of building is most frequently a multi-storied, finished interior property which is a cross between a retail arcade and a loft building.
V. This limitation enables the mart occupants to ship the orders from another location after the retailer or dealer makes his selection from the samples.

The CORRECT answer is:

A. II, IV, III, I, V
B. IV, III, V, I, II
C. I, III, II, IV, V
D. I, IV, II, III, V

5. I. In general, staff-line friction reduces the distinctive contribution of staff personnel. 5.____
II. The conflicts, however, introduce an uncontrolled element into the managerial system.
III. On the other hand, the natural resistance of the line to staff innovations probably usefully restrains over-eager efforts to apply untested procedures on a large scale.
IV. Under such conditions, it is difficult to know when valuable ideas are being sacrificed.
V. The relatively weak position of staff, requiring accommodation to the line, tends to restrict their ability to engage in free, experimental innovation.

The CORRECT answer is:

A. IV, II, III, I, V
B. I, V, III, II, IV
C. V, III, I, II, IV
D. II, I, IV, V, III

KEY (CORRECT ANSWERS)

1. A
2. D
3. D
4. A
5. B

TEST 3

DIRECTIONS: Questions 1 through 4 consist of six sentences which can be arranged in a logical sequence. For each question, select the choice which places the numbered sentences in the MOST logical sequent. *PRINT THE LETTER OF THE CORRECT ANSWER IN THE SPACE AT THE RIGHT.*

1. I. The burden of proof as to each issue is determined before trial and remains upon the same party throughout the trial. 1.____
   II. The jury is at liberty to believe one witness' testimony as against a number of contradictory witnesses.
   III. In a civil case, the party bearing the burden of proof is required to prove his contention by a fair preponderance of the evidence.
   IV. However, it must be noted that a fair preponderance of evidence does not necessarily mean a greater number of witnesses.
   V. The burden of proof is the burden which rests upon one of the parties to an action to persuade the trier of the facts, generally the jury, that a proposition he asserts is true.
   VI. If the evidence is equally balanced, or if it leaves the jury in such doubt as to be unable to decide the controversy either way, judgment must be given against the party upon whom the burden of proof rests.

   The CORRECT answer is:
   A. III, II, V, IV, I, VI
   B. I, II, VI, V, III, IV
   C. III, IV, V, I, II, VI
   D. V, I, III, VI, IV, II

2. I. If a parent is without assets and is unemployed, he cannot be convicted of the crime of non-support of a child. 2.____
   II. The term "sufficient ability" has been held to mean sufficient financial ability.
   III. It does not matter if his unemployment is by choice or unavoidable circumstances.
   IV. If he fails to take any steps at all, he may be liable to prosecution for endangering the welfare of a child.
   V. Under the penal law, a parent is responsible for the support of his minor child only if the parent is "of sufficient ability."
   VI. An indigent parent may meet his obligation by borrowing money or by seeking aid under the provisions of the Social Welfare Law.

   The CORRECT answer is:
   A. VI, I, V, III, II, IV
   B. I, III, V, II, IV, VI
   C. V, II, I, III, VI, IV
   D. I, VI, IV, V, II, III

3. I. Consider, for example, the case of a rabble rouser who urges a group of twenty people to go out and break the windows of a nearby factory. 3.____
   II. Therefore, the law fills the indicated gap with the crime of inciting to riot.
   III. A person is considered guilty of inciting to riot when he urges ten or more persons to engage in tumultuous and violent conduct of a kind likely to create public alarm.
   IV. However, if he has not obtained the cooperation of at least four people, he cannot be charged with unlawful assembly.

V. The charge of inciting to riot was added to the law to cover types of conduct which cannot be classified as either the crime of "riot" or the crime of "unlawful assembly."
VI. If he acquires the acquiescence of at least four of them, he is guilty of unlawful assembly even if the project does not materialize.
The CORRECT answer is:
A. III, V, I, VI, IV, II
B. V, I, IV, VI, II, III
C. III, IV, I, V, II, VI
D. V, I, IV, VI, III, II

4. I. If, however, the rebuttal evidence presents an issue of credibility, it is for the jury to determine whether the presumption has, in fact, been destroyed. 4.____
II. Once sufficient evidence to the contrary is introduced, the presumption disappears from the trial.
III. The effect of a presumption is to place the burden upon the adversary to come forward with evidence to rebut the presumption.
IV. When a presumption is overcome and ceases to exist in the case, the fact or facts which gave rise to the presumption still remain.
V. Whether a presumption has been overcome is ordinarily a question for the court.
VI. Such information may furnish a basis for a logical inference.
The CORRECT answer is:
A. IV, VI, II, V, I, III
B. III, II, V, I, IV, VI
C. V, III, VI, IV, II, I
D. V, IV, I, II, VI, III

KEY (CORRECT ANSWERS)

1. D
2. C
3. A
4. B

PREPARING WRITTEN MATERIALS
EXAMINATION SECTION
TEST 1

DIRECTIONS: Each question or incomplete statement is followed by several suggested answers or completions. Select the one that BEST answers the question or completes the statement. *PRINT THE LETTER OF THE CORRECT ANSWER IN THE SPACE AT THE RIGHT.*

Questions 1-25.

DIRECTIONS: Questions 1 through 25 consist of sentences which may or may not be examples of good English usage. Consider grammar, punctuation, spelling, capitalization, awkwardness, etc. Examine each sentence and then choose the correct statement about it from the four choices below it. If the English usage in the sentence given is better than it would be with any of the changes suggested in options B, C, and D, choose option A. Do not choose an option that will change the meaning of the sentence.

1. According to Judge Frank, the grocer's sons found guilty of assault and sentenced last Thursday. 1.____
   A. This is an example of acceptable writing.
   B. A comma should be placed after the word *sentenced.*
   C. The word *were* should be placed after *sons.*
   D. The apostrophe in grocer's should be placed after the *s.*

2. The department heads assistant said that the stenographers should type duplicate copies of all contracts, leases, and bills. 2.____
   A. This is an example of acceptable writing,
   B. A comma should be placed before the word "*contracts.*
   C. An apostrophe should be placed before the *s* in *heads.*
   D. Quotation marks should be placed before the *stenographers* and after *bills.*

3. The lawyers questioned the men to determine who was the true property owner? 3.____
   A. This is an example of acceptable writing.
   B. The phrase *questioned the men* should be changed to *asked the men questions.*
   C. The word *was* should be changed to *were.*
   D. The question mark should be changed to a period.

4. The terms stated in the present contract are more specific than those stated in the previous contract. 4.____
   A. This is an example of acceptable writing,
   B. The word *are* should be changed to *is*.
   C. The word *than* should be changed to *then*.
   D. The word *specific* should be changed to *specified*.

5. Of the few lawyers considered, the one who argued more skillful was chosen for the job. 5.____
   A. This is an example of acceptable writing.
   B. The word *more* should be replaced by the word *most*.
   C. The word *skillful* should be replaced by the word *skillfully*.
   D. The word *chosen* should be replaced by the word *selected*.

6. Each of the states has a court of appeals; some states have circuit courts. 6.____
   A. This is an example of acceptable writing
   B. The semi-colon should be changed to a comma.
   C. The word *has* should be changed to *have*.
   D. The word *some* should be capitalized.

7. The court trial has greatly effected the child's mental condition. 7.____
   A. This is an example of acceptable writing.
   B. The word *effected* should be changed to *affected*.
   C. The word *greatly* should be placed after *effected*.
   D. The apostrophe in *child's* should be placed after the *s*.

8. Last week, the petition signed by all the officers was sent to the Better Business Bureau. 8.____
   A. This is an example of acceptable writing.
   B. The phrase *last week* should be placed after *officers*.
   C. A comma should be placed after *petition*.
   D. The word *was* should be changed to *were*.

9. Mr. Farrell claims that he requested form A-12, and three booklets describing court procedures. 9.____
   A. This is an example of acceptable writing.
   B. The word *that* should be eliminated.
   C. A colon should be placed after *requested*.
   D. The comma after *A-12* should be eliminated.

10. We attended a staff conference on Wednesday the new safety and fire rules were discussed. 10.____
    A. This is an example of acceptable writing.
    B. The words *safety*, *fire*, and *rules* should begin with capital letters.
    C. There should be a comma after the word *Wednesday*.
    D. There should be a period after the word *Wednesday*, and the word *the* should begin with a capital letter.

11. Neither the dictionary or the telephone directory could be found in the office library. 11.____
    A. This is an example of acceptable writing.
    B. The word *or* should be changed to *nor*.
    C. The word *library* should be spelled *libery*.
    D. The word *neither* should be changed to *either*.

12. The report would have been typed correctly if the typist could read the draft. 12.____
    A. This is an example of acceptable writing.
    B. The word *would* should be removed.
    C. The word *have* should be inserted after the word *could*.
    D. The word *correctly* should be changed to *correct*.

13. The supervisor brought the reports and forms to an employees desk. 13.____
    A. This is an example of acceptable writing.
    B. The word *brought* should be changed to *took*.
    C. There should be a comma after the word *reports* and a comma after the word *forms*.
    D. The word *employees* should be spelled *employee's*.

14. It's important for all the office personnel to submit their vacation schedules on time. 14.____
    A. This is an example of acceptable writing.
    B. The word *It's* should be spelled *Its*.
    C. The word *their* should be spelled *they're*.
    D. The word *personnel* should be spelled *personal*.

15. The supervisor wants that all staff members report to the office at 9:00 A.M. 15.____
    A. This is an example of acceptable writing.
    B. The word *that* should be removed and the word *to* should be inserted after the word *members*.
    C. There should be a comma after the word *wants* and a comma after the word *office*.
    D. The word *wants* should be changed to *want* and the word *shall* should be inserted after the word *members*.

16. Every morning the clerk opens the office mail and distributes it. 16.____
    A. This is an example of acceptable writing.
    B. The word *opens* should be changed to *letters*.
    C. The word *mail* should be changed to *letters*.
    D. The word *it* should be changed to *them*.

17. The secretary typed more fast on a desktop computer than on a tablet. 17.____
    A. This is an example of acceptable writing.
    B. The words *more fast* should be changed to *faster*.
    C. There should be a comma after the words *desktop computer*.
    D. The word *than* should be changed to *then*.

18. The typist used an extention cord in order to connect her typewriter to the outlet nearest to her desks. 18.____
    A. This is an example of acceptable writing.
    B. A period should be placed after the word *cord*, and the word *in* should have a capital *I*.
    C. A comma should be placed after the word *typewriter*.
    D. The word *extention* should be spelled *extension*.

19. He would have went to the conference if he had received an invitation. 19.____
    A. This is an example of acceptable writing.
    B. The word *went* should be replaced by the word *gone*.
    C. The word *had* should be replaced by *would have*.
    D. The word *conference* should be spelled *conferance*.

20. In order to make the report neater, he spent many hours rewriting it. 20.____
    A. This is an example of acceptable writing.
    B. The word *more* should be inserted before the word *neater*.
    C. There should be a colon after the word *neater*,
    D. The word *spent* should be changed to *have spent*.

21. His supervisor told him that he should of read the memorandum more carefully. 21.____
    A. This is an example of acceptable writing.
    B. The word *memorandum* should be spelled *memorandom*.
    C. The word *of* should be replaced by the word *have*.
    D. The word *carefully* should be replaced by the word *careful*.

22. It was decided that two separate reports should be written. 22.____
    A. This is an example of acceptable writing.
    B. A comma should be inserted after the word *decided*.
    C. The word *be* should be replaced by the word *been*.
    D. A colon should be inserted after the word *that*.

23. She don't seem to understand that the work must be done as soon as possible. 23.____
    A. This is an example of acceptable writing.
    B. The word *doesn't* should replace the word *don't*.
    C. The word *why* should replace the word *that*.
    D. The word *as* before the word *soon* should be eliminated.

24. He excepted praise from his supervisor for a job well done. 24.____
    A. This is an example of acceptable writing.
    B. The word *excepted* should be spelled *accepted*.
    C. The order of the words *well done* should be changed to *done well*.
    D. There should be a comma after the word *supervisor*.

25. What appears to be intentional errors in grammar occur several times in the passage. 25.____
   A. This is an example of acceptable writing.
   B. The word *occur* should be spelled *occurr.*
   C. The word *appears* should be changed to *appear.*
   D. The phrase *several times* should be changed to *from time to time.*

KEY (CORRECT ANSWERS)

1.	C	11.	B
2.	C	12.	C
3.	D	13.	D
4.	A	14.	A
5.	C	15.	B
6.	A	16.	A
7.	B	17.	B
8.	A	18.	D
9.	D	19.	B
10.	D	20.	A

21. C
22. A
23. B
24. B
25. C

TEST 2

DIRECTIONS: Each question consists of a sentence which may or may not be an example of good formal English usage. Examine each sentence, considering grammar, punctuation, spelling, capitalization, and awkwardness. Then choose the CORRECT statement about it from the four options below it. If the English usage in the sentence given is better than any of the changes suggested in options B, C, or D, pick option A. Do not pick an option that will change the meaning of the sentence. *PRINT THE LETTER OF THE CORRECT ANSWER IN THE SPACE AT THE RIGHT.*

1. I don't know who could possibly of broken it. 1.____
   A. This is an example of acceptable writing.
   B. The word *who* should be replaced by the word *whom.*
   C. The word *of* should be replaced by the word *have.*
   D. The word *broken* should be replaced by the word *broke.*

2. Telephoning is easier than to write. 2.____
   A. This is an example of acceptable writing.
   B. The word *telephoning* should be spelled *telephoneing.*
   C. The word *than* should be replaced by the word *then.*
   D. The words *to write* should be replaced by the word *writing.*

3. The two operators who have been assigned to these consoles are on vacation. 3.____
   A. This is an example of acceptable writing.
   B. A comma should be placed after the word *operators.*
   C. The word *who* should be replaced by the word *whom.*
   D. The word *are* should be replaced by the word *is.*

4. You were suppose to teach me how to operate a plugboard. 4.____
   A. This is an example of acceptable writing,
   B. The word *were* should be replaced by the word *was.*
   C. The word *suppose* should be replaced by the word *supposed.*
   D. The word *teach* should be replaced by the word *team.*

5. If you had taken my advice; you would have spoken with him. 5.____
   A. This is an example of acceptable writing.
   B. The word *advice* should be spelled *advise.*
   C. The words *had taken* should be replaced by the word *take.*
   D. The semicolon should be changed to a comma.

6. The clerk could have completed the assignment on time if he knows where these materials were located. 6.____
   A. This is an example of acceptable writing.
   B. The word *knows* should be replaced by *had known.*
   C. The word "were" should be replaced by *had been.*
   D. The words *where these materials were located* should be replaced by *the location of these materials.*

7. All employees should be given safety training. Not just those who have accidents. 7.____
   A. This is an example of acceptable writing,
   B. The period after the word *training* should be changed to a colon.
   C. The period after the word *training* should be changed to a semicolon, and the first letter of the word *Not* should be changed to a small *n*.
   D. The period after the word *training* should be changed to a comma, and the first letter of the word *Not* should be changed to a small *n*,

8. This proposal is designed to promote employee awareness of the suggestion program, to encourage employee participation in the program, and to increase the number of suggestions submitted. 8.____
   A. This is an example of acceptable writing.
   B. The word *proposal* should be spelled *proposal*.
   C. The words *to increase the number of suggestions submitted* should be changed to *an increase in the number of suggestions is expected.*
   D. The word *promote* should be changed to *enhance*, and the word *increase* should be changed to *add to.*

9. The introduction of inovative managerial techniques should be preceded by careful analysis of the specific circumstances and conditions in each department. 9.____
   A. This is an example of acceptable writing.
   B. The word *techniques* should be spelled *techneques*.
   C. The word *inovative* should be spelled *innovative*.
   D. A comma should be placed after the word *circumstances* and after the word *conditions.*

10. This occurrence indicates that such criticism embarrasses him. 10.____
    A. This is an example of acceptable writing.
    B. The word *occurrence* should be spelled *occurrence*.
    C. The word *criticism* should be spelled *creticism*.
    D. The word *embarrasses* should be spelled *embarasses*.

11. He can recommend a mechanic whose work is reliable. 11.____
    A. This is an example of acceptable writing.
    B. the word *reliable* should be spelled *relyable*.
    C. The word *whose* should be spelled *who's*.
    D. The word *mechanic* should be spelled *mecanic*.

12. She typed quickly; like someone who had not a moment to lose. 12.____
    A. This is an example of acceptable writing.
    B. The word *not* should be removed.
    C. The semicolon should be changed to a comma.
    D. The word *quickly* should be placed before instead of after the word *typed*.

13. She insisted that she had to much work to do. 13.____
    A. This is an example of acceptable writing.
    B. The word *insisted* should be spelled *insisted.*
    C. The word *to* used in front of *much* should be spelled *too.*
    D. The word *do* should be changed to *be done.*

14. The report, along with the accompanying documents, were submitted for review. 14.____
    A. This is an example of acceptable writing.
    B. The words *were submitted* should be changed to *was submitted.*
    C. The word *accompanying* should be spelled *accompaning.*
    D. The comma after the word *report* should be taken out.

15. If others must use your files, be certain that they understand how the system works, but insist that you do all the filing and refiling. 15.____
    A. This is an example of acceptable writing.
    B. There should be a period after the word *works*, and the word *but* should start a new sentence.
    C. The words *filing* and *refiling* should be spelled *fileing* and *refileing.*
    D. There should be a comma after the word *but.*

16. The appeal was not considered because of its late arrival. 16.____
    A. This is an example of acceptable writing.
    B. The word *its* should be changed to *it's.*
    C. The word *its* should be changed to *the.*
    D. The words *late arrival* should be changed to *arrival late*.

17. The letter must be read carefully to determine under which subject it should be filed. 17.____
    A. This is an example of acceptable writing.
    B. The word *under* should be changed to *at.*
    C. The word *determine* should be spelled *determin.*
    D. The word *carefully* should be spelled *carefully.*

18. He showed potential as an office manager, but he lacked skill in delegating work. 18.____
    A. This is an example of acceptable writing.
    B. The word *delegating* should be spelled *delagating.*
    C. The word *potential* should be spelled *potencial.*
    D. The words *he lacked* should be changed to *was lacking.*

19. His supervisor told him that it would be all right to receive personal mail at the office. 19.____
    A. This is an example of acceptable writing.
    B. The words *all right* should be changed to *alright.*
    C. The word *personal* should be spelled *personel.*
    D. The word *mail* should be changed to *letters.*

20. The report, along with the accompanying documents, were submitted for review. 20.____
    A. This is an example of acceptable writing.
    B. The words *were submitted* should be changed to *was submitted.*
    C. The word *accompanying* should be spelled *accompaning.*
    D. The comma after the word *report* should be taken out.

KEY (CORRECT ANSWERS)

1.	C	11.	A
2.	D	12.	C
3.	A	13.	C
4.	C	14.	B
5.	D	15.	A
6.	B	16.	A
7.	D	17.	D
8.	A	18.	A
9.	C	19.	A
10.	A	20.	B

PHILOSOPHY, PRINCIPLES, PRACTICES, AND TECHNICS
OF
SUPERVISION, ADMINISTRATION, MANAGEMENT, AND ORGANIZATION

TABLE OF CONTENTS

	Page
MEANING OF SUPERVISION	1
THE OLD AND THE NEW SUPERVISION	1
THE EIGHT (8) BASIC PRINCIPLES OF THE NEW SUPERVISION	1
I. Principle of Responsibility	1
II. Principle of Authority	2
III. Principle of Self-Growth	2
IV. Principle of Individual Worth	2
V. Principle of Creative Leadership	2
VI. Principle of Success and Failure	2
VII. Principle of Science	3
VIII. Principle of Cooperation	3
WHAT IS ADMINISTRATION?	3
I. Practices Commonly Classed as "Supervisory"	3
II. Practices Commonly Classed as "Administrative"	3
III. Practices Commonly Classed as Both "Supervisory" and "Administrative"	4
RESPONSIBILITIES OF THE SUPERVISOR	4
COMPETENCIES OF THE SUPERVISOR	4
THE PROFESSIONAL SUPERVISOR-EMPLOYEE RELATIONSHIP	4
MINI-TEXT IN SUPERVISION, ADMINISTRATION, MANAGEMENT, AND ORGANIZATION	5
I. Brief Highlights	5
A. Levels of Management	6
B. What the Supervisor Must Learn	6
C. A Definition of Supervision	6
D. Elements of the Team Concept	6
E. Principles of Organization	6
F. The Four Important Parts of Every Job	7
G. Principles of Delegation	7
H. Principles of Effective Communications	7
I. Principles of Work Improvement	7
J. Areas of Job Improvement	7
K. Seven Key Points in Making Improvements	8

L. Corrective Techniques for Job Improvement 8
M. A Planning Checklist 8
N. Five Characteristics of Good Directions 9
O. Types of Directions 9
P. Controls 9
Q. Orienting the New Employee 9
R. Checklist for Orienting New Employees 9
S. Principles of Learning 10
T. Causes of Poor Performance 10
U. Four Major Steps in On-the-Job Instructions 10
V. Employees Want Five Things 10
W. Some Don'ts in Regard to Praise 11
X. How to Gain Your Workers' Confidence 11
Y. Sources of Employee Problems 11
Z. The Supervisor's Key to Discipline 11
AA. Five Important Processes of Management 12
BB. When the Supervisor Fails to Plan 12
CC. Fourteen General Principles of Management 12
DD. Change 12

II. Brief Topical Summaries 13
A. Who/What is the Supervisor? 13
B. The Sociology of Work 13
C. Principles and Practices of Supervision 14
D. Dynamic Leadership 14
E. Processes for Solving Problems 15
F. Training for Results 15
G. Health, Safety, and Accident Prevention 16
H. Equal Employment Opportunity 16
I. Improving Communications 16
J. Self-Development 17
K. Teaching and Training 17
1. The Teaching Process 17
a. Preparation 17
b. Presentation 18
c. Summary 18
d. Application 18
e. Evaluation 18
2. Teaching Methods 18
a. Lecture 18
b. Discussion 18
c. Demonstration 19
d. Performance 19
e. Which Method to Use 19

PHILOSOPHY, PRINCIPLES, PRACTICES, AND TECHNICS
OF
SUPERVISION, ADMINISTRATION, MANAGEMENT, AND ORGANIZATION

MEANING OF SUPERVISION

The extension of the democratic philosophy has been accompanied by an extension in the scope of supervision. Modern leaders and supervisors no longer think of supervision in the narrow sense of being confined chiefly to visiting employees, supplying materials, or rating the staff. They regard supervision as being intimately related to all the concerned agencies of society, they speak of the supervisor's function in terms of "growth," rather than the "improvement" of employees.

This modern concept of supervision may be defined as follows: Supervision is leadership and the development of leadership within groups which are cooperatively engaged in inspection, research, training, guidance, and evaluation.

THE OLD AND THE NEW SUPERVISION

TRADITIONAL

1. Inspection
2. Focused on the employee
3. Visitation
4. Random and haphazard
5. Imposed and authoritarian
6. One person usually

MODERN

1. Study and analysis
2. Focused on aims, materials, methods, supervisors, employees, environment
3. Demonstrations, intervisitation, workshops, directed reading, bulletins, etc.
4. Definitely organized and planned (scientific)
5. Cooperative and democratic
6. Many persons involved (creative)

THE EIGHT (8) BASIC PRINCIPLES OF THE NEW SUPERVISION

I. Principle of Responsibility
Authority to act and responsibility for acting must be joined.
A. If you give responsibility, give authority.
B. Define employee duties clearly.
C. Protect employees from criticism by others.
D. Recognize the rights as well as obligations of employees.
E. Achieve the aims of a democratic society insofar as it is possible within the area of your work.
F. Establish a situation favorable to training and learning.
G. Accept ultimate responsibility for everything done in your section, unit, office, division, department.
H. Good administration and good supervision are inseparable.

II. Principle of Authority

The success of the supervisor is measured by the extent to which the power of authority is not used.

A. Exercise simplicity and informality in supervision
B. Use the simplest machinery of supervision
C. If it is good for the organization as a whole, it is probably justified.
D. Seldom be arbitrary or authoritative.
E. Do not base your work on the power of position or of personality.
F. Permit and encourage the free expression of opinions.

III. Principle of Self-Growth

The success of the supervisor is measured by the extent to which, and the speed with which, he is no longer needed.

A. Base criticism on principles, not on specifics.
B. Point out higher activities to employees.
C. Train for self-thinking by employees to meet new situations.
D. Stimulate initiative, self-reliance, and individual responsibility
E. Concentrate on stimulating the growth of employees rather than on removing defects.

IV. Principle of Individual Worth

Respect for the individual is a paramount consideration in supervision.

A. Be human and sympathetic in dealing with employees.
B. Don't nag about things to be done.
C. Recognize the individual differences among employees and seek opportunities to permit best expression of each personality.

V. Principle of Creative Leadership

The best supervision is that which is not apparent to the employee.

A. Stimulate, don't drive employees to creative action.
B. Emphasize doing good things.
C. Encourage employees to do what they do best.
D. Do not be too greatly concerned with details of subject or method.
E. Do not be concerned exclusively with immediate problems and activities.
F. Reveal higher activities and make them both desired and maximally possible.
G. Determine procedures in the light of each situation but see that these are derived from a sound basic philosophy.
H. Aid, inspire, and lead so as to liberate the creative spirit latent in all good employees.

VI. Principle of Success and Failure

There are no unsuccessful employees, only unsuccessful supervisors who have failed to give proper leadership.

A. Adapt suggestions to the capacities, attitudes, and prejudices of employees.
B. Be gradual, be progressive, be persistent.
C. Help the employee find the general principle; have the employee apply his own problem to the general principle.
D. Give adequate appreciation for good work and honest effort.
E. Anticipate employee difficulties and help to prevent them.
F. Encourage employees to do the desirable things they will do anyway.
G. Judge your supervision by the results it secures.

VII. Principle of Science
Successful supervision is scientific, objective, and experimental. It is based on facts, not on prejudices.
A. Be cumulative in results.
B. Never divorce your suggestions from the goals of training.
C. Don't be impatient of results.
D. Keep all matters on a professional, not a personal, level.
E. Do not be concerned exclusively with immediate problems and activities.
F. Use objective means of determining achievement and rating where possible.

VIII. Principle of Cooperation
Supervision is a cooperative enterprise between supervisor and employee.
A. Begin with conditions as they are.
B. Ask opinions of all involved when formulating policies.
C. Organization is as good as its weakest link.
D. Let employees help to determine policies and department programs.
E. Be approachable and accessible—physically and mentally.
F. Develop pleasant social relationships.

WHAT IS ADMINISTRATION

Administration is concerned with providing the environment, the material facilities, and the operational procedures that will promote the maximum growth and development of supervisors and employees. (Organization is an aspect and a concomitant of administration.)

There is no sharp line of demarcation between supervision and administration; these functions are intimately interrelated and, often, overlapping. They are complementary activities.

I. Practices Commonly Classed as "Supervisory"
A. Conducting employees' conferences
B. Visiting sections, units, offices, divisions, departments
C. Arranging for demonstrations
D. Examining plans
E. Suggesting professional reading
F. Interpreting bulletins
G. Recommending in-service training courses
H. Encouraging experimentation
I. Appraising employee morale
J. Providing for intervisitation

II. Practices Commonly Classified as "Administrative"
A. Management of the office
B. Arrangement of schedules for extra duties
C. Assignment of rooms or areas
D. Distribution of supplies
E. Keeping records and reports
F. Care of audio-visual materials
G. Keeping inventory records
H. Checking record cards and books

I. Programming special activities
J. Checking on the attendance and punctuality of employees

III. Practices Commonly Classified as Both "Supervisory" and "Administrative"
A. Program construction
B. Testing or evaluating outcomes
C. Personnel accounting
D. Ordering instructional materials

RESPONSIBILITIES OF THE SUPERVISOR

A person employed in a supervisory capacity must constantly be able to improve his own efficiency and ability. He represent the employer to the employees and only continuous self-examination can make him a capable supervisor.

Leadership and training are the supervisor's responsibility. An efficient working unit is one in which the employees work with the supervisor. It is his job to bring out the best in his employees. He must always be relaxed, courteous, and calm in his association with his employees. Their feelings are important, and a harsh attitude does not develop the most efficient employees.

COMPETENCES OF THE SUPERVISOR

I. Complete knowledge of the duties and responsibilities of his position.
II. To be able to organize a job, plan ahead, and carry through.
III. To have self-confidence and initiative.
IV. To be able to handle the unexpected situation and make quick decisions.
V. To be able to properly train subordinates in the positions they are best suited for.
VI. To be able to keep good human relations among his subordinates.
VII. To be able to keep good human relations between his subordinates and himself and to earn their respect and trust.

THE PROFESSIONAL SUPERVISOR-EMPLOYEE RELATIONSHIP

There are two kinds of efficiency: one kind is only apparent and is produced in organizations through the exercise of mere discipline; this is but a simulation of the second, or true, efficiency which springs from spontaneous cooperation. If you are a manager, no matter how great or small your responsibility, it is your job, in the final analysis, to create and develop this involuntary cooperation among the people whom you supervise. For, no matter how powerful a combination of money, machines, and materials a company may have, this is a dead and sterile thing without a team of willing, thinking, and articulate people to guide it.

The following 21 points are presented as indicative of the exemplary basic relationship that should exist between supervisor and employee:

1. Each person wants to be liked and respected by his fellow employee and wants to be treated with consideration and respect by his superior.
2. The most competent employee will make an error. However, in a unit where good relations exist between the supervisor and his employees, tenseness and fear do not exist. Thus, errors are not hidden or covered up, and the efficiency of a unit is not impaired.

3. Subordinates resent rules, regulations, or orders that are unreasonable or unexplained.
4. Subordinates are quick to resent unfairness, harshness, injustices, and favoritism.
5. An employee will accept responsibility if he knows that he will be complimented for a job well done, and not too harshly chastised for failure; that his supervisor will check the cause of the failure, and, if it was the supervisor's fault, he will assume the blame therefore. If it was the employee's fault, his supervisor will explain the correct method or means of handling the responsibility.
6. An employee wants to receive credit for a suggestion he has made, that is used. If a suggestion cannot be used, the employee is entitled to an explanation. The supervisor should not say "no" and close the subject.
7. Fear and worry slow up a worker's ability. Poor working environment can impair his physical and mental health. A good supervisor avoids forceful methods, threats, and arguments to get a job done.
8. A forceful supervisor is able to train his employees individually and as a team, and is able to motivate them in the proper channels.
9. A mature supervisor is able to properly evaluate his subordinates and to keep them happy and satisfied.
10. A sensitive supervisor will never patronize his subordinates.
11. A worthy supervisor will respect his employees' confidences.
12. Definite and clear-cut responsibilities should be assigned to each executive.
13. Responsibility should always be coupled with corresponding authority.
14. No change should be made in the scope or responsibilities of a position without a definite understanding to that effect on the part of all persons concerned.
15. No executive or employee, occupying a single position in the organization, should be subject to definite orders from more than one source.
16. Orders should never be given to subordinates over the head of a responsible executive. Rather than do this, the officer in question should be supplanted.
17. Criticisms of subordinates should, whoever possible, be made privately, and in no case should a subordinate be criticized in the presence of executives or employees of equal or lower rank.
18. No dispute or difference between executives or employees as to authority or responsibilities should be considered too trivial for prompt and careful adjudication.
19. Promotions, wage changes, and disciplinary action should always be approved by the executive immediately superior to the one directly responsible.
20. No executive or employee should ever be required, or expected, to be at the same time an assistant to, and critic of, another.
21. Any executive whose work is subject to regular inspection should, wherever practicable, be given the assistance and facilities necessary to enable him to maintain an independent check of the quality of his work.

MINI-TEXT IN SUPERVISION, ADMINISTRATION, MANAGEMENT, AND ORGANIZATION

I. Brief Highlights

Listed concisely and sequentially are major headings and important data in the field for quick recall and review.

A. Levels of Management

Any organization of some size has several levels of management. In terms of a ladder, the levels are:

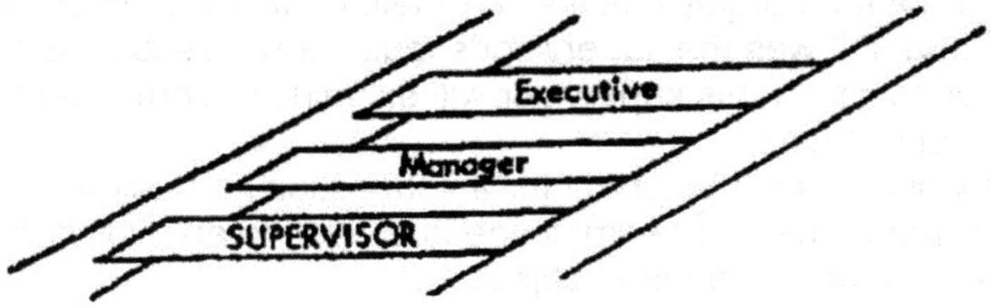

The first level is very important because it is the beginning point of management leadership.

B. What the Supervisor Must Learn

A supervisor must learn to:

1. Deal with people and their differences
2. Get the job done through people
3. Recognize the problems when they exist
4. Overcome obstacles to good performance
5. Evaluate the performance of people
6. Check his own performance in terms of accomplishment

C. A Definition of Supervisor

The term supervisor means any individual having authority, in the interests of the employer, to hire, transfer, suspend, lay-off, recall, promote, discharge, assign, reward, or discipline other employees or responsibility to direct them, or to adjust their grievances, or effectively to recommend such action, if, in connection with the foregoing, exercise of such authority is not of a merely routine or clerical nature but requires the use of independent judgment.

D. Elements of the Team Concept

What is involved in teamwork? The component parts are:

1. Members
2. A leader
3. Goals
4. Plans
5. Cooperation
6. Spirit

E. Principles of Organization

1. A team member must know what his job is.
2. Be sure that the nature and scope of a job are understood.
3. Authority and responsibility should be carefully spelled out.
4. A supervisor should be permitted to make the maximum number of decisions affecting his employees.
5. Employees should report to only one supervisor.
6. A supervisor should direct only as many employees as he can handle effectively.
7. An organization plan should be flexible.

8. Inspection and performance of work should be separate.
9. Organizational problems should receive immediate attention.
10. Assign work in line with ability and experience.

F. The Four Important Parts of Every Job
1. Inherent in every job is the *accountability* for results.
2. A second set of factors in every job is *responsibilities*.
3. Along with duties and responsibilities one must have the *authority* to act within certain limits without obtaining permission to proceed.
4. No job exists in a vacuum. The supervisor is surrounded by key *relationships*.

G. Principles of Delegation

Where work is delegated for the first time, the supervisor should think in terms of these questions:
1. Who is best qualified to do this?
2. Can an employee improve his abilities by doing this?
3. How long should an employee spend on this?
4. Are there any special problems for which he will need guidance?
5. How broad a delegation can I make?

H. Principles of Effective Communications
1. Determine the media.
2. To whom directed?
3. Identification and source authority.
4. Is communication understood?

I. Principles of Work Improvement
1. Most people usually do only the work which is assigned to them.
2. Workers are likely to fit assigned work into the time available to perform it.
3. A good workload usually stimulates output.
4. People usually do their best work when they know that results will be reviewed or inspected.
5. Employees usually feel that someone else is responsible for conditions of work, workplace layout, job methods, type of tools/equipment, and other such factors.
6. Employees are usually defensive about their job security.
7. Employees have natural resistance to change.
8. Employees can support or destroy a supervisor.
9. A supervisor usually earns the respect of his people through his personal example of diligence and efficiency.

J. Areas of Job Improvement

The areas of job improvement are quite numerous, but the most common ones which a supervisor can identify and utilize are:
1. Departmental layout
2. Flow of work
3. Workplace layout
4. Utilization of manpower
5. Work methods
6. Materials handling

7. Utilization
8. Motion economy

K. Seven Key Points in Making Improvements
1. Select the job to be improved
2. Study how it is being done now
3. Question the present method
4. Determine actions to be taken
5. Chart proposed method
6. Get approval and apply
7. Solicit worker participation

I. Corrective Techniques of Job Improvement

Specific Problems
1. Size of workload
2. Inability to meet schedules
3. Strain and fatigue
4. Improper use of men and skills
5. Waste, poor quality, unsafe conditions
6. Bottleneck conditions that hinder output
7. Poor utilization of equipment and machine
8. Efficiency and productivity of labor

General Improvement
1. Departmental layout
2. Flow of work
3. Work plan layout
4. Utilization of manpower
5. Work methods
6. Materials handling
7. Utilization of equipment
8. Motion economy

Corrective Techniques
1. Study with scale model
2. Flow chart study
3. Motion analysis
4. Comparison of units produced to standard allowance
5. Methods analysis
6. Flow chart and equipment study
7. Down time vs. running time
8. Motion analysis

M. A Planning Checklist
1. Objectives
2. Controls
3. Delegations
4. Communications
5. Resources
6. Manpower

7. Equipment
8. Supplies and materials
9. Utilization of time
10. Safety
11. Money
12. Work
13. Timing of improvements

N. Five Characteristics of Good Directions
In order to get results, directions must be:
1. Possible of accomplishment
2. Agreeable with worker interests
3. Related to mission
4. Planned and complete
5. Unmistakably clear

O. Types of Directions
1. Demands or direct orders
2. Requests
3. Suggestion or implication
4. volunteering

P. Controls
A typical listing of the overall areas in which the supervisor should establish controls might be:
1. Manpower
2. Materials
3. Quality of work
4. Quantity of work
5. Time
6. Space
7. Money
8. Methods

Q. Orienting the New Employee
1. Prepare for him
2. Welcome the new employee
3. Orientation for the job
4. Follow-up

R. Checklist for Orienting New Employees

	Yes	No
1. Do you appreciate the feelings of new employees when they first report for work?	___	___
2. Are you aware of the fact that the new employee must make a big adjustment to his job?	___	___
3. Have you given him good reasons for liking the job and the organization?	___	___
4. Have you prepared for his first day on the job?	___	___
5. Did you welcome him cordially and make him feel needed?	___	___

		Yes	No
6.	Did you establish rapport with him so that he feels free to talk and discuss matters with you?	___	___
7.	Did you explain his job to him and his relationship to you?	___	___
8.	Does he know that his work will be evaluated periodically on a basis that is fair and objective?	___	___
9.	Did you introduce him to his fellow workers in such a way that they are likely to accept him?	___	___
10.	Does he know what employee benefits he will receive?	___	___
11.	Does he understand the importance of being on the job and what to do if he must leave his duty station?	___	___
12.	Has he been impressed with the importance of accident prevention and safe practice?	___	___
13.	Does he generally know his way around the department?	___	___
14.	Is he under the guidance of a sponsor who will teach the right way of doing things?	___	___
15.	Do you plan to follow-up so that he will continue to adjust successfully to his job?	___	___

S. Principles of Learning
1. Motivation
2. Demonstration or explanation
3. Practice

T. Causes of Poor Performance
1. Improper training for job
2. Wrong tools
3. Inadequate directions
4. Lack of supervisory follow-up
5. Poor communications
6. Lack of standards of performance
7. Wrong work habits
8. Low morale
9. Other

U. Four Major Steps in On-The-Job Instruction
1. Prepare the worker
2. Present the operation
3. Tryout performance
4. Follow-up

V. Employees Want Five Things
1. Security
2. Opportunity
3. Recognition
4. Inclusion
5. Expression

W. Some Don'ts in Regard to Praise
1. Don't praise a person for something he hasn't done.
2. Don't praise a person unless you can be sincere.
3. Don't be sparing in praise just because your superior withholds it from you.
4. Don't let too much time elapse between good performance and recognition of it

X. How to Gain Your Workers' Confidence
Methods of developing confidence include such things as:
1. Knowing the interests, habits, hobbies of employees
2. Admitting your own inadequacies
3. Sharing and telling of confidence in others
4. Supporting people when they are in trouble
5. Delegating matters that can be well handled
6. Being frank and straightforward about problems and working conditions
7. Encouraging others to bring their problems to you
8. Taking action on problems which impede worker progress

Y. Sources of Employee Problems
On-the-job causes might be such things as:
1. A feeling that favoritism is exercised in assignments
2. Assignment of overtime
3. An undue amount of supervision
4. Changing methods or systems
5. Stealing of ideas or trade secrets
6. Lack of interest in job
7. Threat of reduction in force
8. Ignorance or lack of communications
9. Poor equipment
10. Lack of knowing how supervisor feels toward employee
11. Shift assignments

Off-the-job problems might have to do with:
1. Health
2. Finances
3. Housing
4. Family

Z. The Supervisor's Key to Discipline
There are several key points about discipline which the supervisor should keep in mind:
1. Job discipline is one of the disciplines of life and is directed by the supervisor.
2. It is more important to correct an employee fault than to fix blame for it.
3. Employee performance is affected by problems both on the job and off.
4. Sudden or abrupt changes in behavior can be indications of important employee problems.
5. Problems should be dealt with as soon as possible after they are identified.
6. The attitude of the supervisor may have more to do with solving problems than the techniques of problem solving.
7. Correction of employee behavior should be resorted to only after the supervisor is sure that training or counseling will not be helpful.

8. Be sure to document your disciplinary actions.
9. Make sure that you are disciplining on the basis of facts rather than personal feelings.
10. Take each disciplinary step in order, being careful not to make snap judgments, or decisions based on impatience.

AA. Five Important Processes of Management

1. Planning
2. Organizing
3. Scheduling
4. Controlling
5. Motivating

BB. When the Supervisor Fails to Plan

1. Supervisor creates impression of not knowing his job
2. May lead to excessive overtime
3. Job runs itself—supervisor lacks control
4. Deadlines and appointments missed
5. Parts of the work go undone
6. Work interrupted by emergencies
7. Sets a bad example
8. Uneven workload creates peaks and valleys
9. Too much time on minor details at expense of more important tasks

CC. Fourteen General Principles of Management

1. Division of work
2. Authority and responsibility
3. Discipline
4. Unity of command
5. Unity of direction
6. Subordination of individual interest to general interest
7. Remuneration of personnel
8. Centralization
9. Scalar chain
10. Order
11. Equity
12. Stability of tenure of personnel
13. Initiative
14. Esprit de corps

DD. Change

Bringing about change is perhaps attempted more often, and yet less well understood, than anything else the supervisor does. How do people generally react to change? (People tend to resist change that is imposed upon them by other individuals or circumstances.

Change is characteristic of every situation. It is a part of every real endeavor where the efforts of people are concerned.

1. Why do people resist change?
   People may resist change because of:
   a. Fear of the unknown
   b. Implied criticism
   c. Unpleasant experiences in the past
   d. Fear of loss of status
   e. Threat to the ego
   f. Fear of loss of economic stability

2. How can we best overcome the resistance to change?
   In initiating change, take these steps:
   a. Get ready to sell
   b. Identify sources of help
   c. Anticipate objections
   d. Sell benefits
   e. Listen in depth
   f. Follow up

II. Brief Topical Summaries

A. Who/What is the Supervisor?

1. The supervisor is often called the "highest level employee and the lowest level manager."
2. A supervisor is a member of both management and the work group. He acts as a bridge between the two.
3. Most problems in supervision are in the area of human relations, or people problems.
4. Employees expect: Respect, opportunity to learn and to advance, and a sense of belonging, and so forth.
5. Supervisors are responsible for directing people and organizing work. Planning is of paramount importance.
6. A position description is a set of duties and responsibilities inherent to a given position.
7. It is important to keep the position description up-to-date and to provide each employee with his own copy.

B. The Sociology of Work

1. People are alike in many ways; however, each individual is unique.
2. The supervisor is challenged in getting to know employee differences. Acquiring skills in evaluating individuals is an asset.
3. Maintaining meaningful working relationships in the organization is of great importance.
4. The supervisor has an obligation to help individuals to develop to their fullest potential.
5. Job rotation on a planned basis helps to build versatility and to maintain interest and enthusiasm in work groups.
6. Cross training (job rotation) provides backup skills.

7. The supervisor can help reduce tension by maintaining a sense of humor, providing guidance to employees, and by making reasonable and timely decisions. Employees respond favorably to working under reasonably predictable circumstances.
8. Change is characteristic of all managerial behavior. The supervisor must adjust to changes in procedures, new methods, technological changes, and to a number of new and sometimes challenging situations.
9. To overcome the natural tendency for people to resist change, the supervisor should become more skillful in initiating change.

C. Principles and Practices of Supervision

1. Employees should be required to answer to only one superior.
2. A supervisor can effectively direct only a limited number of employees, depending upon the complexity, variety, and proximity of the jobs involved.
3. The organizational chart presents the organization in graphic form. It reflects lines of authority and responsibility as well as interrelationships of units within the organization.
4. Distribution of work can be improved through an analysis using the "Work Distribution Chart."
5. The "Work Distribution Chart" reflects the division of work within a unit in understandable form.
6. When related tasks are given to an employee, he has a better chance of increasing his skills through training.
7. The individual who is given the responsibility for tasks must also be given the appropriate authority to insure adequate results.
8. The supervisor should delegate repetitive, routine work. Preparation of recurring reports, maintaining leave and attendance records are some examples.
9. Good discipline is essential to good task performance. Discipline is reflected in the actions of employees on the job in the absence of supervision.
10. Disciplinary action may have to be taken when the positive aspects of discipline have failed. Reprimand, warning, and suspension are examples of disciplinary action.
11. If a situation calls for a reprimand, be sure it is deserved and remember it is to be done in private.

D. Dynamic Leadership

1. A style is a personal method or manner of exerting influence.
2. Authoritarian leaders often see themselves as the source of power and authority.
3. The democratic leader often perceives the group as the source of authority and power.
4. Supervisors tend to do better when using the pattern of leadership that is most natural for them.
5. Social scientists suggest that the effective supervisor use the leadership style that best fits the problem or circumstances involved.
6. All four styles—telling, selling, consulting, joining—have their place. Using one does not preclude using the other at another time.

7. The theory X point of view assumes that the average person dislikes work, will avoid it whenever possible, and must be coerced to achieve organizational objectives.
8. The theory Y point of view assumes that the average person considers work to be a natural as play, and, when the individual is committed, he requires little supervision or direction to accomplish desired objectives.
9. The leader's basic assumptions concerning human behavior and human nature affect his actions, decisions, and other managerial practices.
10. Dissatisfaction among employees is often present, but difficult to isolate. The supervisor should seek to weaken dissatisfaction by keeping promises, being sincere and considerate, keeping employees informed, and so forth.
11. Constructive suggestions should be encouraged during the natural progress of the work.

E. Processes for Solving Problems

1. People find their daily tasks more meaningful and satisfying when they can improve them.
2. The causes of problems, or the key factors, are often hidden in the background. Ability to solve problems often involves the ability to isolate them from their backgrounds. There is some substance to the cliché that some persons "can't see the forest for the trees."
3. New procedures are often developed from old ones. Problems should be broken down into manageable parts. New ideas can be adapted from old one.
4. People think differently in problem-solving situations. Using a logical, patterned approach is often useful. One approach found to be useful includes these steps:
   a. Define the problem
   b. Establish objectives
   c. Get the facts
   d. Weigh and decide
   e. Take action
   f. Evaluate action

F. Training for Results

1. Participants respond best when they feel training is important to them.
2. The supervisor has responsibility for the training and development of those who report to him.
3. When training is delegated to others, great care must be exercised to insure the trainer has knowledge, aptitude, and interest for his work as a trainer.
4. Training (learning) of some type goes on continually. The most successful supervisor makes certain the learning contributes in a productive manner to operational goals.
5. New employees are particularly susceptible to training. Older employees facing new job situations require specific training, as well as having need for development and growth opportunities.
6. Training needs require continuous monitoring.
7. The training officer of an agency is a professional with a responsibility to assist supervisors in solving training problems.

8. Many of the self-development steps important to the supervisor's own growth are equally important to the development of peers and subordinates. Knowledge of these is important when the supervisor consults with others on development and growth opportunities.

G. Health, Safety, and Accident Prevention

1. Management-minded supervisors take appropriate measures to assist employees in maintaining health and in assuring safe practices in the work environment.
2. Effective safety training and practices help to avoid injury and accidents.
3. Safety should be a management goal. All infractions of safety which are observed should be corrected without exception.
4. Employees' safety attitude, training and instruction, provision of safe tools and equipment, supervision, and leadership are considered highly important factors which contribute to safety and which can be influenced directly by supervisors.
5. When accidents do occur, they should be investigated promptly for very important reasons, including the fact that information which is gained can be used to prevent accidents in the future.

H. Equal Employment Opportunity

1. The supervisor should endeavor to treat all employees fairly, without regard to religion, race, sex, or national origin.
2. Groups tend to reflect the attitude of the leader. Prejudice can be detected even in very subtle form. Supervisors must strive to create a feeling of mutual respect and confidence in every employee.
3. Complete utilization of all human resources is a national goal. Equitable consideration should be accorded women in the work force, minority-group members, the physically and mentally handicapped, and the older employee. The important question is: "Who can do the job?"
4. Training opportunities, recognition for performance, overtime assignments, promotional opportunities, and all other personnel actions are to be handled on an equitable basis.

I. Improving Communications

1. Communications is achieving understanding between the sender and the receiver of a message. It also means sharing information—the creation of understanding.
2. Communication is basic to all human activity. Words are means of conveying meanings; however, real meanings are in people.
3. There are very practical differences in the effectiveness of one-way, impersonal, and two-way communications. Words spoken face-to-face are better understood. Telephone conversations are effective, but lack the rapport of person-to-person exchanges. The whole person communicates.
4. Cooperation and communication in an organization go hand in hand. When there is a mutual respect between people, spelling out rules and procedures for communicating is unnecessary.
5. There are several barriers to effective communications. These include failure to listen with respect and understanding, lack of skill in feedback, and misinterpreting the meanings of words used by the speaker. It is also common

practice to listen to what we want to hear, and tune out things we do not want to hear.

6. Communication is management's chief problem. The supervisor should accept the challenge to communicate more effectively and to improve interagency and intra-agency communications.
7. The supervisor may often plan for and conduct meetings. The planning phase is critical and may determine the success or the failure of a meeting.
8. Speaking before groups usually requires extra effort. Stage fright may never disappear completely, but it can be controlled.

J. Self-Development

1. Every employee is responsible for his own self-development.
2. Toastmaster and toastmistress clubs offer opportunities to improve skills in oral communications.
3. Planning for one's own self-development is of vital importance. Supervisors know their own strengths and limitations better than anyone else.
4. Many opportunities are open to aid the supervisor in his developmental efforts, including job assignments; training opportunities, both governmental and non-governmental—to include universities and professional conferences and seminars.
5. Programmed instruction offers a means of studying at one's own rate.
6. Where difficulties may arise from a supervisor's being away from his work for training, he may participate in televised home study or correspondence courses to meet his self-development needs.

K. Teaching and Training

1. The Teaching Process

Teaching is encouraging and guiding the learning activities of students toward established goals. In most cases this process consists of five steps: preparation, presentation, summarization, evaluation, and application.

a. Preparation

Preparation is two-fold in nature; that of the supervisor and the employee. Preparation by the supervisor is absolutely essential to success. He must know what, when, where, how, and whom he will teach. Some of the factors that should be considered are:

1) The objectives
2) The materials needed
3) The methods to be used
4) Employee participation
5) Employee interest
6) Training aids
7) Evaluation
8) Summarization

Employee preparation consists in preparing the employee to receive the material. Probably the most important single factor in the preparation of the employee is arousing and maintaining his interest. He must know the objectives of the training, why he is there, how the material can be used, and its importance to him.

b. Presentation
In presentation, have a carefully designed plan and follow it. The plan should be accurate and complete, yet flexible enough to meet situations as they arise. The method of presentation will be determined by the particular situation and objectives.

c. Summary
A summary should be made at the end of every training unit and program. In addition, there may be internal summaries depending on the nature of the material being taught. The important thing is that the trainee must always be able to understand how each part of the new material relates to the whole.

d. Application
The supervisor must arrange work so the employee will be given a chance to apply new knowledge or skills while the material is still clear in his mind and interest is high. The trainee does not really know whether he has learned the material until he has been given a chance to apply it. If the material is not applied, it loses most of its value.

e. Evaluation
The purpose of all training is to promote learning. To determine whether the training has been a success or failure, the supervisor must evaluate this learning.
In the broadest sense, evaluation includes all the devices, methods, skills, and techniques used by the supervisor to keep himself and the employees informed as to their progress toward the objectives they are pursuing. The extent to which the employee has mastered the knowledge, skills, and abilities, or changed his attitudes, as determined by the program objectives, is the extent to which instruction has succeeded or failed.
Evaluation should not be confined to the end of the lesson, day, or program but should be used continuously. We shall note later the way this relates to the rest of the teaching process.

2. Teaching Methods
A teaching method is a pattern of identifiable student and instructor activity used in presenting training material.
All supervisors are faced with the problem of deciding which method should be used at a given time.

a. Lecture
The lecture is direct oral presentation of material by the supervisor. The present trend is to place less emphasis on the trainer's activity and more on that of the trainee.

b. Discussion
Teaching by discussion or conference involves using questions and other techniques to arouse interest and focus attention upon certain areas, and by doing so creating a learning situation. This can be one of the most

valuable methods because it gives the employees an opportunity to express their ideas and pool their knowledge.

c. Demonstration
The demonstration is used to teach how something works or how to do something. It can be used to show a principle or what the results of a series of actions will be. A well-staged demonstration is particularly effective because it shows proper methods of performance in a realistic manner.

d. Performance
Performance is one of the most fundamental of all learning techniques or teaching methods. The trainee may be able to tell how a specific operation should be performed but he cannot be sure he knows how to perform the operation until he has done so.
As with all methods, there are certain advantages and disadvantages to each method.

e. Which Method to Use
Moreover, there are other methods and techniques of teaching. It is difficult to use any method without other methods entering into it. In any learning situation, a combination of methods is usually more effective than any one method alone.

Finally, evaluation must be integrated into the other aspects of the teaching-learning process.

It must be used in the motivation of the trainees; it must be used to assist in developing understanding during the training; and it must be related to employee application of the results of training.

This is distinctly the role of the supervisor.

ACCIDENT DIAGRAMS

One section of the exam will test your ability to understand accident diagrams. Each question presents a description of an accident. In some questions, your understanding of directions (north, south, east, and west) will be tested. In other questions, you must choose the diagram that BEST represents the description of an accident.

Symbols are used to represent vehicles and pedestrians and their movements. These symbols and their meaning will be used in the test.

Moving vehicles are represented by this symbol: front  rear

Parked vehicles are represented by this symbol: front  rear

Pedestrians and the direction in which they are heading are represented by a circle and an arrow: 

Bicyclists and the direction in which they are heading are represented by this symbol and an arrow: 

Solid lines indicate the path and direction of a vehicle or person *before* an accident happens: 

Dotted lines indicate the path and direction of a vehicle or person *after* an accident happens: 

SAMPLE QUESTIONS

Question 1 is a diagram of an accident. You are to determine the missing directions. Read the question, study the diagram, and then choose the set of directions that matches the diagram of the accident.

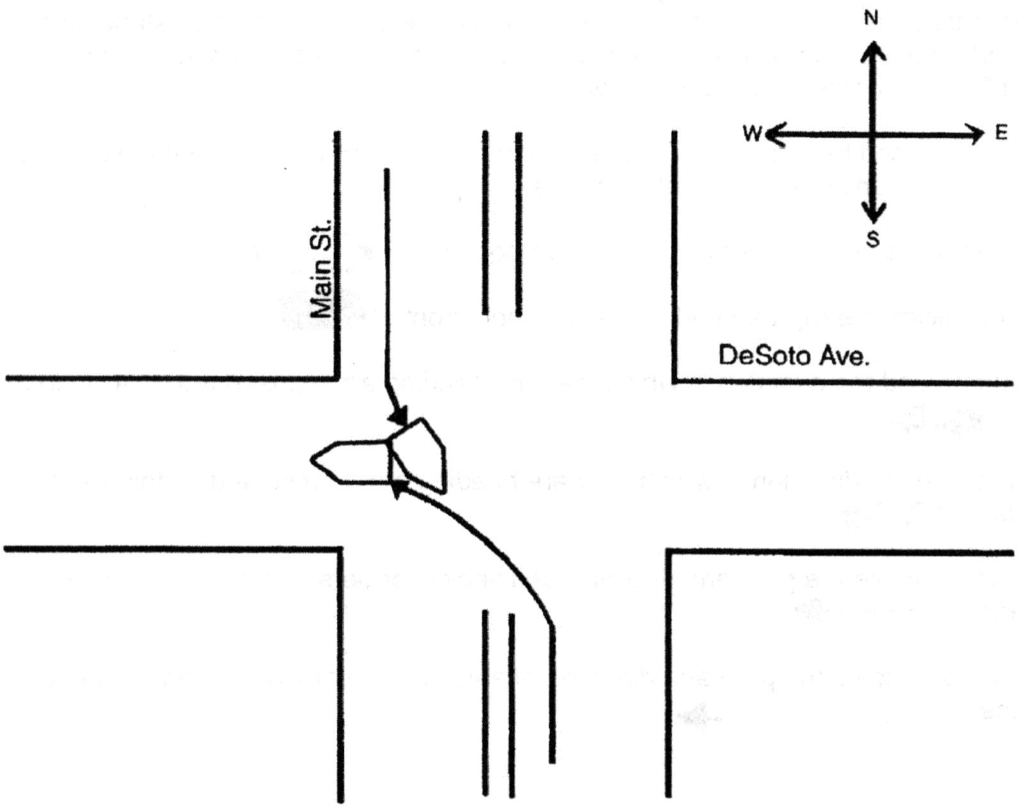

1. A car traveling north turns _____ onto DeSoto Avenue and swerves to avoid a second car heading _____ on Main Street attempting to make a _____ turn heading _____ on DeSoto. 1._____
   A. right; south, right; west
   B. left; south; left; east
   C. left; north; left; east
   D. right; north; right; east

Question 2 contains a description of an accident. Choose the diagram that BEST represents the accident.

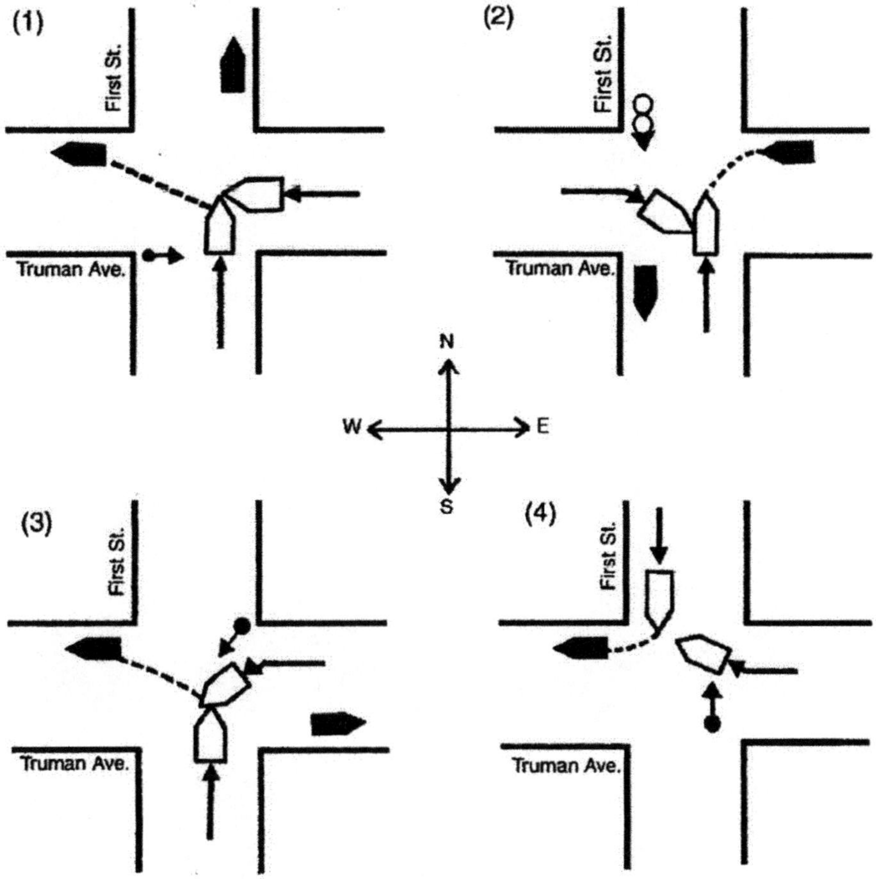

2. A car heading west on Truman Avenue swerved to avoid hitting a pedestrian and was hit by a second car heading north on First Street. The first car was pushed into a car parked on the north side of Truman Avenue. 2.____
Which of the four diagrams above BEST illustrates the accident?
A. 1 B. 2 C. 3 D. 4

KEY (CORRECT ANSWERS)

1. B
2. C

TRAFFIC CONTROL STUDIES

CONTENTS

	Page
Traffic Control Device Studies	1
Vehicle Registration Study	1
Origin-Destination Study	1
Speed Study	1
Speed-Delay Study	1
Motor Vehicle Volume Study	2
Roadway Capacity Studies	2
Vehicle Occupancy Study	2
Pedestrian Study	2
Observance of Stop Sign Study	3
Observance of Traffic Signals Study	3
Parking Studies	3
Accident Records Study	3

TRAFFIC CONTROL STUDIES

Type	Purpose	Requirement For Study	Personnel and Equipment
Traffic Control Device Studies	To inventory, locate, classify, and evaluate traffic control devices; and increase adequacy of these devices.	One initial study of all devices which is updated by periodic studies of specific areas on a routine basis.	Special two-man teams. Normal patrol equipment, and stopwatch, tape measure (100 ft), manual on uniform traffic control devices, field forms or notebook.
Vehicle Registration Study	To determine peak loads of traffic and adequacy of parking. May be used to adjust or update origin and destination study, or be used in lieu of this study.	As required to measure peak traffic in relation to existing roadways, and duty hour schedules.	Study is conducted by extraction and processing of information with ADPS. Traffic section personnel obtain input data, and ADP section processes data as required.
Origin-Destination Study	To develop data on origins and destinations of personnel entering, leaving, or traveling within an installation on a typical working day.	As required to support long-range planning, to anticipate major changes in strengths and functions, to support traffic construction requirements, and to assign traffic properly.	Varies with type and scope of study.
Speed Study	To determine if prevailing speeds are proper, to determine proper speed for new or improve roadways; to serve as a warrant for, and guide in, the placement and operation of traffic control devices, and to assist in accident research and enforcement.	Conducted for specific roadways as a result of observation, enforcement activity, and accident experience. Also required for new or renovated roadways.	Personnel may consist of one-man or two-man teams depending on the method and type of study. Equipment may consist of patrol vehicle, mirror box, stopwatch, field sheets, radar (with or without graphic recorder), and electric timer. Normally, police gear and marked vehicles are not used.
Speed-Delay Study	To determine variation in speed along a route; indicate amount, location, course, frequency and duration of delays, and provide overall speed and travel time along a route.	Conducted on specific routes as problems develop of congestion, delay and insufficient capacity. Also conducted when necessary to assign route priority, to consider use of alternate routes, to evaluate speed limits, and to check effectiveness of control devices.	Personnel will consist of a two-man team without distinctive police gear. Unmarked sedan or ¼-ton truck, standard watch and stopwatch, and field sheets as required.

Type	Purpose	Requirement For Study	Personnel and Equipment
Motor Vehicle Volume Study	To obtain an accurate record of the number, directional movements, and variation in volume of motor vehicles passing through intersections or using major routes, and to provide data for use in construction of a traffic flow map.	Conducted as required to determine street adequacy, to appraise effectiveness of traffic control measures, and to establish priorities and designs for traffic and/or road improvements and for new streets.	Two policemen are required to observe and record at a normal two-way intersection. If traffic exceeds 1500 vehicles per hour entering the intersection, one policeman may be required for each of the four approaches. Ordinary watches, field sheets, summary sheets, and (if used) manual counters are needed.
Roadway Capacity Studies	To determine the practical capacities of roadways as an adjunct to other studies; and to provide basic information required to update traffic regulations, to establish priorities for street improvements, and to aid in traffic planning.	Conducted as required to relieve congestion through appropriate corrective action in those areas where traffic volumes exceed traffic capacities.	Varies with scope of study. Normally, as a minimum, requires a two-man team equipped with tape measure, stopwatch post or engineer maps, sketch pads, and odometer (optional).
Vehicle Occupancy Study	To determine the number of occupants per motor vehicle.	As required to examine parking difficulties and congestion; to assist in planning for future traffic and parking facilities, and to evaluate the adequacy of transit services.	Either one-man or two-man teams with normal police gear depending on traffic volume. Equipment required includes ordinary watch, field sheets, and summary sheets.
Pedestrian Study	To determine the amount of pedestrian traffic at intersections and/or midblock crossing points.	As required to evaluate pedestrian-vehicle conflicts, and assist in planning control, physical protection, and enforcement measures.	Locally designed field sheets or notebooks. Either one-man or two-man teams depending on the pedestrian volume. Police gear is not worn.
Observance of Stop Sign Study	To determine the degree of drive obedience.	As required to study the relation of driver obedience to accidents at high accident frequency locations, and to assist in taking measures to increase driver obedience.	One person can normally make this study. He should not wear distinctive police equipment and should have a watch and field sheets.

Type	Purpose	Requirement For Study	Personnel and Equipment
Observance of Traffic Signals Study	To determine voluntary observance of intersection traffic control signals.	As required at intersections where congestion and high accident rates prevail.	Two policemen without distinctive police gear are normally required. On multiple approaches with heavy traffic, four or six policemen may be required. Equipment consists of an ordinary watch, field sheets, and summary sheets.
Parking Studies	To determine the adequacy, use, and location of existing parking facilities; and to provide guidance in the placement and design of parking areas for future use.	A comprehensive, installation survey is normally required only in conjunction with long-range planning for major changes in the installation. Surveys are conducted at specific areas as parking problems become evident, or in anticipation of the development of parking problems.	Field sheets, summary sheets, post map, aerial photos, and questionnaires are used as required for the specific study or survey being conducted. Personnel requirements and use of police gear depend on the type and scope of the study.
Accident Records Study	To improve enforcement, engineering, and education programs.	As needed to identify and treat high accident locations, to assist in evaluating highway design factors, to establish priorities of action, and to measure effectiveness of remedial action.	ADP equipment and trained personnel for automatic data processing. Normally, two police perform observations for condition and collision diagrams.

GLOSSARY OF TRAFFIC CONTROL TERMS

TABLE OF CONTENTS

	Page
Access Road ... Desire Line	1
Divided Street ... Left Turn Lane	2
Manual Traffic Control ... Passenger Vehicle	3
Passenger (Transit) Volume ... Separate Turning Lane	4
Shoulder ... Traffic Accident	5
Traffic Actuated Controller ... Uninterrupted Flow	6
Vehicle ... Zone (Origin-Destination Studies)	7

GLOSSARY OF TRAFFIC CONTROL TERMS

A

ACCESS ROAD - Public roads, existing or proposed, needed to provide essential access to military installation and facilities, or to industrial installations and facilities in the activities of which there is specific defense interest. Roads within the boundaries of military reservation are excluded from this definition unless such roads have been dedicated to public use and are not subject to closure.

ACCIDENT SPOT MAP - An area or installation map showing the location of vehicle accidents by means of symbols. Symbols may represent accidents classified as to daylight hours, night hours, injury or death.

ANGLE PARKING - Parking where the longitudinal axes of vehicles form an angle with the alignment of the roadway.

C

CENTER LINE - A line marking the center of a roadway between traffic moving in opposite direction.

COLLISION DIAGRAM - A plan of an intersection or section of roadway on which reported accidents are diagramed by means of arrows showing manner of collision.

COMBINED CONDITION AND COLLISION DIAGRAM - A condition diagram upon which the reported accidents are diagramed by means of arrows showing manner of collision.

CONDITION DIAGRAM - A plan of an intersection or section of roadway showing all objects and physical conditions having a bearing on traffic movement and safety at that location. Usually these are scaled drawings.

CORDON COUNTS - A count of all vehicles and persons entering and leaving a district (cordon area) during a designated period of time.

CORDON AREA - The district bounded by the cordon line and included in a cordon count.

CROSSWALK - Any portion of a roadway at an intersection or elsewhere distinctly indicated for pedestrian crossing by lines or other markings on the surface. Also, that part of a roadway at an intersection included within the connections of the lateral lines of the sidewalks on opposite sides of the traffic way measured from the curbs, or in the absence of curbs, from the edges of the traversable roadway.

D

DELAY - The time consumed while traffic or a specified component of traffic is impeded in its movement by some element over which it has no control usually expressed in seconds per vehicle.

DESIRE LINE - A straight line between the point of origin and point of destination of a trip without regard to routes of travel (used in connection with an origin-destination study).

DIVIDED STREET - A two-way road on which traffic in one direction of travel is separated from that in the opposite direction by a directional separator. Such a road has two or more roadways.

E

85 PERCENTILE SPEED - That speed below which 85 percent of the traffic unit's travel, and above which 15 percent travel.

F

FIXED-TIME CONTROLLER - An automatic controller for supervising the operation of traffic control signals in accordance with a predetermined fixed time cycle and divisions thereof.

FIXED-TIME TRAFFIC SIGNAL - A traffic signal operated by a fixed-time controller.

FLASHING BEACON - A section of a standard traffic signal head, or a similar type device, having a yellow or red lens in each face, which is illuminated by rapid intermittent flashes.

FLASHING TRAFFIC SIGNAL - A traffic control signal used as a flashing beacon.

FLOATING CAR- An automobile driven in the traffic flow at the average speed of the surrounding vehicles.

FLOW DIAGRAM - The graphical representation of the traffic volumes on a road or street network or section thereof, showing by means of bands the relative volumes using each section of roadway during a given period of time, usually 1 hour.

H

HIGH FREQUENCY ACCIDENT LOCATION - A specific location where a large number of traffic accidents have occurred.

I

INTERSECTION APPROACH - That portion of an intersection leg which is used by traffic approaching the intersection.

L

LATERAL CLEARANCE - The distance between the edge of pavement and any lateral obstruction.

LATERAL OBSTRUCTION- Any fixed object located adjacent to the traveled way which reduces the transverse dimensions of the roadway.

LEFT TURN LANE - A lane within the normal surfaced width reserved for left turning vehicles.

3

M

MANUAL TRAFFIC CONTROL - The use of-hand signals or manually operated devices by traffic control personnel to control traffic.

MANUAL COUNTER - A tallying device which is operated by hand.

MASS TRANSPORTATION - Movement of large groups of persons.

MULTIAXLE TRUCK - A truck which has more than two axles.

O

OCCUPANCY RATIO -The average number-of occupants per vehicle (including the driver).

ODOMETER -A device on a vehicle for measuring the distance traveled, usually as a cumulative total, but sometimes also for individual trips, with an indicator on the instrument panel where it is usually combined with a speedometer indicator, or in the hub of a wheel in some trucks.

OFF-PEAK PERIOD - That portion of the day in which traffic volumes are relatively light.

OFFSET LANES - Additional lanes used for traffic which is heavier in one direction. Also known as unbalanced lanes.

OFF-STREET PARKING - Lots and garages intended for parking entirely off streets and alleys. street and alleys (may be angle or parallel parking) for parking of vehicles.

ORIGIN DESTINATION STUDIES - A study of the origins and destinations of trips of vehicles and passengers. Usually included in the study are all trips within, or passing through, into or out of a selected area.

OVERALL SPEED - The total distance traversed divided by the travel time. Usually expressed in miles per hour and includes all delays.

OVERALL TIME - The time of travel, including stops and delays except those off the traveled way.

P

PARALLEL PARKING - Parking where the longitudinal axis of vehicles are parallel to alignment of the roadway so that the vehicles are facing in the same direction as the movement of adjacent vehicular traffic.

PARKING DURATION - Length of time a vehicle is parked.

PASSENGER VEHICLE - A free-wheeled, self-propelled vehicle designed for the transportation of persons but limited in seating capacity to not more than seven passengers, not including the driver. It includes taxicabs, limousines, and station wagons, but does not include motorcycles.
(In capacity studies, also includes light reconnaissance vehicles, and pickup trucks.)

PASSENGER (TRANSIT) VOLUME - The total number of public transit occupants being transported in a period of time.

PEAK PERIOD - That portion of the day in which maximum traffic volumes are experienced.

PEDESTRIAN - Any person afoot. For purpose of accident classification, this will be interpreted to include any person riding in or upon a device moved or designed for movement by human power or the force of gravity, except bicycles, including stilts, skates, skis, sleds, toy wagons, and scooters.

PERCENT OF GRADE - The slope in the longitudinal direction of the pavement expressed in percent which is the number of units of change in elevation per 100 units of horizontal distance.

PERCENT OF GREEN TIME - The percentage of green time allotted to the direction of travel being studies.

PROPERTY DAMAGE - Damage to property as a result of a motor vehicle accident that may be a basis of a claim for compensation. Does not include compensation for loss of life or for personal injuries.

PUBLIC HIGHWAYS- The entire width between property lines, or boundary lines, of every way or place of which any part is open to use of the public for purposes of vehicular traffic as a matter of right or custom.

PUBLIC TRANSIT - The public passenger carryi ng service afforded by vehicles following regular routes and making specified stops.

R

REFLECTORIZE - The application of some material to traffic control devices or hazards which will return to the eyes of the road user some portion of the light from his vehicle headlights, thereby producing a brightness which attracts attention.

REGULATORY DEVICE - A device used to indicate the required method of traffic movement or use of the public traffic way.

REGULATORY SIGN - A sign used to indicate the required method of traffic movement or use of the traffic way.

RIGHT TURN LANE - A lane within the normal surfaced width reserved for right turning vehicles.

ROADWAY - That portion of a traffic way including shoulders, improved, designed, or ordinarily used for vehicle traffic.

S

SEPARATE TURNING LANE - Added traffic lane which is separated from the intersection area by an island or unpaved area. It may be wide enough for one or two line operation

SHOULDER - The portion of the roadway contiguous with the traveled way for accommodation of stopped vehicles, for emergency use, and for lateral support of base and surface courses.

SIGHT DISTANCES - The length of roadway visible to the driver of a passenger vehicle at any given point on the roadway when the view is unobstructed by traffic.

SIGNAL CYCLE - The total time required for one complete sequence of the intervals of a traffic signal.

SIGNAL CONTROLLER - A complete electrical mechanism for controlling the operation of traffic control signals, including the timer and all necessary auxiliary apparatus mounted in a cabinet.

SIGNAL FACE - That part of a signal head provided for controlling traffic from a single direction.

SIGNAL HEAD - An assembly containing one or more signal faces that may be designated accordingly as one-way, two-way, multi-way.

SIGNAL PHASE - A part of the total time cycle allocated to movements receiving the right-of-way or to any combination ments receiving the right-of-way simultaneously during one

SIMPLE INTERSECTION - An intersection of two traffic ways, approaches.

SPEED - The rate of movement of a vehicle, generally expressed in miles per hour.

STOPPING SIGHT DISTANCE –The distance required by a drive of a vehicle, given speed, to bring vehicle to a stop after and object becomes visible.

STREET WIDTH - The width of the paved or traveled portion of the roadway.

T

THROUGH MOVEMENT - (See THROUGH TRAFFIC)

THROUGH STREET - A street on which traffic is given the right-of-way so that vehicles entering or crossing the street must yield the right-of-way.

THROUGH TRAFFIC - Traffic proceeding through a military installation or portion not originating in or destined to that military installation or portion thereof.

TIME CYCLE - (See SIGNAL CYCLE)

TRAFFIC - Pedestrians, ridden or herded animals, vehicles, street cars, and other conveyances, either singly or together, while using any street for purposes of travel.

TRAFFIC ACCIDENT - Any accident involving a motor vehicle in motion that results in death, injury, or property damage.

6

TRAFFIC ACTUATED CONTROLLER- An automatic controller for supervising the operation of traffic control signals in accordance with the immediate and varying demands of traffic as registered with the-controller by means of detectors.

TRAFFIC CONTROL - All measures except those of a structural kind that serve to control and guide traffic and to promote road safety.

TRAFFIC CONTROL DEVICE - A Traffic control device is any sign, signal, marking, or device placed or erected for the purpose of regulating, warning, or guiding traffic.

TRAFFIC DEMAND - The volume of traffic desiring to use a particular route or facility.

TRAFFIC ENGINEERING - That phase of engineering that deals with the planning and geometric design of streets, highways, and abutting lands, and with traffic operations thereon, as their use is related to the safe, convenient, and economic transportation of persons and goods.

TRAFFIC FLOW - The movement of vehicles on a roadway.

TRAFFIC FLOW PATTERN - The distribution of traffic volumes on a street or highway network~

TRAFFIC GENERATOR - A traffic producing area such as a post exchange, parking lot, or administrative center.

TRAFFIC SIGNAL INTERVAL - Anyone of the several divisions of the total time cycle during which signal indications do not change.

TRAFFICWAY - The entire width between property lines (or other boundary lines) of every way or place of which any part is open to use of public for purposes of vehicular traffic as a matter of right or custom.

TRANSIT VEHICLE - A passenger carrying vehicle, such as a bus or streetcar which follows regular routes and makes specific stops.

TRAVEL TIME- The total elapsed time from the origin to destination of a trip.

TURNING MOVEMENT - The traffic making a designated turn at an intersection.

TWO-WAY STREETS - A street on which traffic may move in opposite directions simultaneously. It may be either divided or undivided.

TYPE OF ACCIDENT - The kind of motor vehicle accident, such as head-on, right-angle, etc.

TYPE OF SURFACE - The class of surface such as concrete, asphalt, gravel, etc.

U

UNINTERRRUPTED FLOW - The flow of-vehicles under ideal conditions resulting in unrestricted movement.

7

V

VEHICLE - Every device in, upon, or by which any person or property is or may be transported or drawn upon a highway, except those devices moved by human power or used exclusively upon stationary rails or tracks.

VEHICULE OCCUPANCY - The average number of occupants per automobile, including the driver.

VOLUME - The number of vehicles passing a given point during a specified period of time.

W

WARNING SIGN - A sign used to indicate conditions that are actually or potentially hazardous to highway users.

WARRANT - Formally stated conditions that have been accepted as minimum requirements for justifying installation of a traffic control device or regulation.

Z

ZONE (ORIGIN-DESTINATION STUDIES) -- A division of an area established for the purpose of analyzing origin-destination studies. It may be bounded by physical barriers such as rivers and highways, or may be the location of individual work organizations that have duty stations in relatively close proximity.

ANSWER SHEET

TEST NO. ____________ PART ________ TITLE OF POSITION ______________________________
(AS GIVEN IN EXAMINATION ANNOUNCEMENT - INCLUDE OPTION, IF ANY)

PLACE OF EXAMINATION ______________________ (CITY OR TOWN) ______________ (STATE) DATE __________

RATING

USE THE SPECIAL PENCIL. MAKE GLOSSY BLACK MARKS.

1	A B C D E	26	A B C D E	51	A B C D E	76	A B C D E	101	A B C D E
2	A B C D E	27	A B C D E	52	A B C D E	77	A B C D E	102	A B C D E
3	A B C D E	28	A B C D E	53	A B C D E	78	A B C D E	103	A B C D E
4	A B C D E	29	A B C D E	54	A B C D E	79	A B C D E	104	A B C D E
5	A B C D E	30	A B C D E	55	A B C D E	80	A B C D E	105	A B C D E
6	A B C D E	31	A B C D E	56	A B C D E	81	A B C D E	106	A B C D E
7	A B C D E	32	A B C D E	57	A B C D E	82	A B C D E	107	A B C D E
8	A B C D E	33	A B C D E	58	A B C D E	83	A B C D E	108	A B C D E
9	A B C D E	34	A B C D E	59	A B C D E	84	A B C D E	109	A B C D E
10	A B C D E	35	A B C D E	60	A B C D E	85	A B C D E	110	A B C D E

Make only ONE mark for each answer. Additional and stray marks may be counted as mistakes. In making corrections, erase errors COMPLETELY.

11	A B C D E	36	A B C D E	61	A B C D E	86	A B C D E	111	A B C D E
12	A B C D E	37	A B C D E	62	A B C D E	87	A B C D E	112	A B C D E
13	A B C D E	38	A B C D E	63	A B C D E	88	A B C D E	113	A B C D E
14	A B C D E	39	A B C D E	64	A B C D E	89	A B C D E	114	A B C D E
15	A B C D E	40	A B C D E	65	A B C D E	90	A B C D E	115	A B C D E
16	A B C D E	41	A B C D E	66	A B C D E	91	A B C D E	116	A B C D E
17	A B C D E	42	A B C D E	67	A B C D E	92	A B C D E	117	A B C D E
18	A B C D E	43	A B C D E	68	A B C D E	93	A B C D E	118	A B C D E
19	A B C D E	44	A B C D E	69	A B C D E	94	A B C D E	119	A B C D E
20	A B C D E	45	A B C D E	70	A B C D E	95	A B C D E	120	A B C D E
21	A B C D E	46	A B C D E	71	A B C D E	96	A B C D E	121	A B C D E
22	A B C D E	47	A B C D E	72	A B C D E	97	A B C D E	122	A B C D E
23	A B C D E	48	A B C D E	73	A B C D E	98	A B C D E	123	A B C D E
24	A B C D E	49	A B C D E	74	A B C D E	99	A B C D E	124	A B C D E
25	A B C D E	50	A B C D E	75	A B C D E	100	A B C D E	125	A B C D E

ANSWER SHEET

TEST NO. ________ PART ______ TITLE OF POSITION ____________________
(AS GIVEN IN EXAMINATION ANNOUNCEMENT - INCLUDE OPTION, IF ANY)

PLACE OF EXAMINATION ____________________ (CITY OR TOWN) ____________ (STATE) DATE ______

RATING

USE THE SPECIAL PENCIL. MAKE GLOSSY BLACK MARKS.

1 A B C D E	26 A B C D E	51 A B C D E	76 A B C D E	101 A B C D E
2 A B C D E	27 A B C D E	52 A B C D E	77 A B C D E	102 A B C D E
3 A B C D E	28 A B C D E	53 A B C D E	78 A B C D E	103 A B C D E
4 A B C D E	29 A B C D E	54 A B C D E	79 A B C D E	104 A B C D E
5 A B C D E	30 A B C D E	55 A B C D E	80 A B C D E	105 A B C D E
6 A B C D E	31 A B C D E	56 A B C D E	81 A B C D E	106 A B C D E
7 A B C D E	32 A B C D E	57 A B C D E	82 A B C D E	107 A B C D E
8 A B C D E	33 A B C D E	58 A B C D E	83 A B C D E	108 A B C D E
9 A B C D E	34 A B C D E	59 A B C D E	84 A B C D E	109 A B C D E
10 A B C D E	35 A B C D E	60 A B C D E	85 A B C D E	110 A B C D E

Make only ONE mark for each answer. Additional and stray marks may be counted as mistakes. In making corrections, erase errors COMPLETELY.

11 A B C D E	36 A B C D E	61 A B C D E	86 A B C D E	111 A B C D E
12 A B C D E	37 A B C D E	62 A B C D E	87 A B C D E	112 A B C D E
13 A B C D E	38 A B C D E	63 A B C D E	88 A B C D E	113 A B C D E
14 A B C D E	39 A B C D E	64 A B C D E	89 A B C D E	114 A B C D E
15 A B C D E	40 A B C D E	65 A B C D E	90 A B C D E	115 A B C D E
16 A B C D E	41 A B C D E	66 A B C D E	91 A B C D E	116 A B C D E
17 A B C D E	42 A B C D E	67 A B C D E	92 A B C D E	117 A B C D E
18 A B C D E	43 A B C D E	68 A B C D E	93 A B C D E	118 A B C D E
19 A B C D E	44 A B C D E	69 A B C D E	94 A B C D E	119 A B C D E
20 A B C D E	45 A B C D E	70 A B C D E	95 A B C D E	120 A B C D E
21 A B C D E	46 A B C D E	71 A B C D E	96 A B C D E	121 A B C D E
22 A B C D E	47 A B C D E	72 A B C D E	97 A B C D E	122 A B C D E
23 A B C D E	48 A B C D E	73 A B C D E	98 A B C D E	123 A B C D E
24 A B C D E	49 A B C D E	74 A B C D E	99 A B C D E	124 A B C D E
25 A B C D E	50 A B C D E	75 A B C D E	100 A B C D E	125 A B C D E